Advances in Particle Swarm Optimization

Advances in Particle Swarm Optimization

Editor: May Church

STATES
ACADEMIC PRESS
www.statesacademicpress.com

Published by States Academic Press
109 South 5th Street,
Brooklyn, NY 11249, USA
www.statesacademicpress.com

Advances in Particle Swarm Optimization
Edited by May Church

International Standard Book Number: 978-1-63989-024-8 (Hardback)

Cataloging-in-Publication Data

Advances in particle swarm optimization / edited by May Church.
p. cm.
Includes bibliographical references and index.
ISBN 978-1-63989-024-8
1. Mathematical optimization. 2. Particles (Nuclear physics).
3. Swarm intelligence. I. Church, May.
QA402.5 .A38 2022
519.3--dc23

Contents

Preface

This book has been a concerted effort by a group of academicians, researchers and scientists, who have contributed their research works for the realization of the book. This book has materialized in the wake of emerging advancements and innovations in this field. Therefore, the need of the hour was to compile all the required researches and disseminate the knowledge to a broad spectrum of people comprising of students, researchers and specialists of the field.

Particle swarm optimization can be defined as a computational method that is used to optimize a problem by iteratively trying to improve a candidate solution with respect to a given measure of quality. It is deployed to solve a problem by having a population of candidate solutions and moving them around in the search-space in accordance with simple mathematical formulae over the particle's position and velocity. Particle swarm optimization can search very large spaces of candidate solutions because it is metaheuristic and does not make any assumptions about the problem being optimized. There are various variants of particle swamp optimization such as hybridization, simplifications, multi-objective optimization, and binary, discrete, and combinational particle swamp optimization. This book elucidates the concepts and innovative models around prospective developments in relation to particle swarm optimization. Different approaches, evaluations, methodologies, and advanced studies on this topic have been included in it. This book will serve as a reference to a broad spectrum of readers.

At the end of the preface, I would like to thank the authors for their brilliant chapters and the publisher for guiding us all-through the making of the book till its final stage. Also, I would like to thank my family for providing the support and encouragement throughout my academic career and research projects.

Editor

1

Searching for the Best Points of Interpolation using Swarm Intelligence Techniques

Djerou L.[1], Khelil N.[1], Zerarka A.[1] and Batouche M.[2]
[1]*Labo de Mathématiques Appliquées, Université Med Khider*
[2]*Computer Science Department, CCIS-King Saud University*
[1]*Algeria,* [2]*Saudi Arabia*

1. Introduction

If the values of a function, $f(x)$, are known for a finite set of x values in a given interval, then a polynomial which takes on the same values at these x values offers a particularly simple analytic approximation to $f(x)$ throughout the interval. This approximating technique is called polynomial interpolation. Its effectiveness depends on the smoothness of the function being sampled (if the function is unknown, some hypothetical smoothness must be chosen), on the number and choice of points at which the function is sampled.

In practice interpolating polynomials with degrees greater than about 10 are rarely used. One of the major problems with polynomials of high degree is that they tend to oscillate wildly. This is clear if they have many roots in the interpolation interval. For example, a degree 10 polynomial with 10 real roots must cross the x-axis 10 times. Thus, it would not be suitable for interpolating a monotone decreasing or increasing function on such an interval.

In this chapter we explore the advantage of using the Particle Swarm Optimization (PSO) interpolation nodes. Our goal is to show that the PSO nodes can approximate functions with much less error than Chebyshev nodes.

This chapter is organized as follows. In Section 2, we shall present the interpolation polynomial in the Lagrange form. Section 3 examines the Runge's phenomenon; which illustrates the error that can occur when constructing a polynomial interpolant of high degree. Section 4 gives an overview of modern heuristic optimization techniques, including fundamentals of computational intelligence for PSO. We calculate in Subsection 4.2 the best interpolating points generated by PSO algorithm. We make in section 5, a comparison of interpolation methods. The comments and conclusion are made in Section 6.

2. Introduction to the Lagrange interpolation

If $x_0, x_1, \ldots.x_n$ are distinct real numbers, then for arbitrary values $y_0, y_1, \ldots.y_n$, there is a unique polynomial p_n of degree at most n such that $p_n(x_i) = y_i$ $(0 \leq i \leq n)$ (David Kincaid & Ward Cheney, 2002).

The Lagrange form looks as follows:

$$p_n\left(x\right)=y_0l_0\left(x\right)+\ldots+y_nl_n\left(x\right)=\sum_{i=0}^{n}y_il_i\left(x\right) \tag{1}$$

Such that coordinal functions can be expressed in the following

$$l_i\left(x\right)=\prod_{j=0,j\neq i}^{n}\frac{\left(x-x_j\right)}{\left(x_i-x_j\right)} \tag{2}$$

(David Kincaid & Ward Cheney, 2002), (Roland E. et al 1994).
are coordinal polynomials that satisfy

$$l_i(x_j)=\begin{cases}1, & i=j\\ 0, & i\neq j\end{cases} \tag{3}$$

The Lagrange form gives an error term of the form

$$E_n\left(x\right)=f(x)-p_n\left(x\right)=\frac{f^{(n+1)}\left(c\right)}{\left(n+1\right)!}\phi_n\left(x\right) \tag{4}$$

Where

$$\phi_n\left(x\right)=\prod_{i=0}^{n}\left(x-x_i\right) \tag{5}$$

If we examine the error formula for polynomial interpolation over an interval [a, b] we see that as we change the interpolation points, we change also the locations c where the derivative is evaluated; thus that part in the error also changes, and that change is a "black hole" to us: we never know what the correct value of c is, but only that c is somewhere in the interval [a, b]. Since we wish to use the interpolating polynomial to approximate the Equation (4) cannot be used, of course, to calculate the exact value of the error f – P_n, since c, as a function of x is, in general, not known. (An exception occurs when the (n + 1)st derivative off is constant). And so we are likely to reduce the error by selecting interpolation points $x_0, x_1, \ldots, x_n$ so as to minimize the maximum value of product $\phi_n\left(x\right)$

The most natural idea is to choose them regularly distribute in [a, b].

3. Introduction to the Runge phenomenon and to Chebyshev approximations

3.1 Runge phenomenon

If x_k are chosen to be the points $x_k=a+k\dfrac{b-a}{n}\quad\left(k=0,\ldots,n\right)$ (means that are equally spaced at a distance $2n + 1$ apart), then the interpolating polynomial $p_n(x)$ need not to converge uniformly on [a, b] as $n \longrightarrow \infty$ for the function f(x).

This phenomenon is known as the Runge phenomenon (RP) and it can be illustrated with the Runge's "bell" function on the interval [-5, 5] (Fig.1).

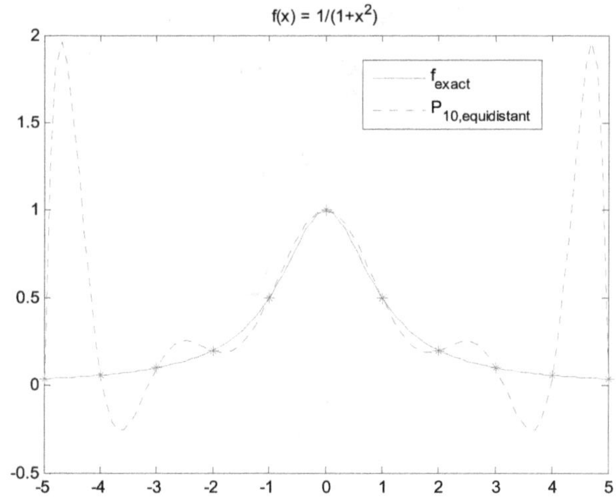

Figure 1. solid blue line: present Runge's "bell" function. dots red line: present the polynomial approximation based on equally 11 spaced nodes

3.2 Chebyshev Nodes

The standard remedy against the RP is Chebyshev -type clustering of nodes towards the end of the interval (Fig.3).

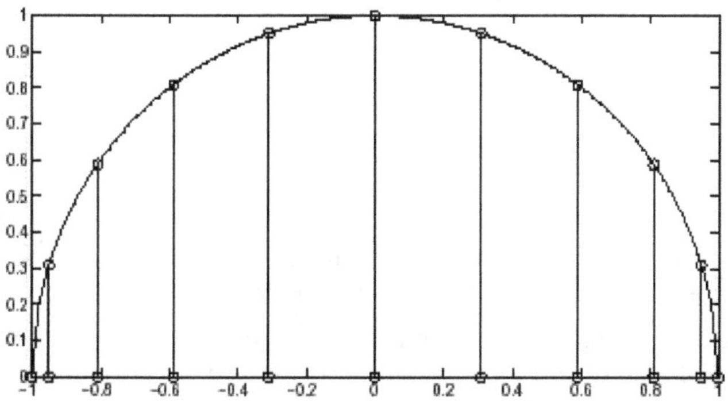

Figure 2. Chebyshev Point Distribution.

To do this, conceptually, we would like to take many points near the endpoints of the interval and few near the middle. The point distribution that minimizes the maximum value of product $\phi_n(x)$ is called the Chebyshev distribution, as shown in (Fig. 2). In the Chebyshev distribution, we proceed as follows:

1. Draw the semicircle on [a, b].

2. To sample $n + 1$ points, place n equidistant partitions on the arc.
3. Project each partition onto the x-axis: for j =0, 1… n

$$x_j = \frac{a+b}{2} + \frac{b-a}{2}\cos\left(j\frac{\pi}{n}\right) \text{ for } j=0,1\ldots n \tag{6}$$

The nodes x_i that will be used in our approximation are:

Chebyshev nodes
-5.0000
-4.2900
-4.0251
-2.6500
-1.4000
0.0000
1.4000
2.6500
4.0451
4.2900
5.0000

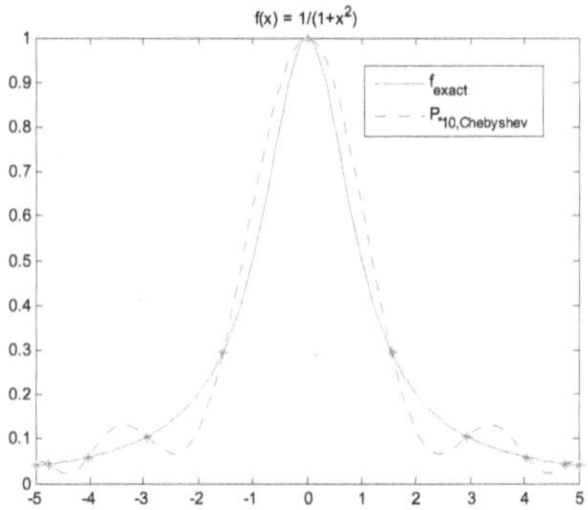

Figure 3. solid blue line: present Runge's "bell" function. dots red line: present the polynomial approximation based on 11 Chebyshev nodes

In this study, we have made some numerical computations using the particle swarm optimization to investigate the best interpolating points and we are showing that PSO nodes provide smaller approximation error than Chebyshev nodes.

4. Particle swarm optimization

4.1 Overview and strategy of particle swarm optimization

Recently, a new stochastic algorithm has appeared, namely 'particle swarm optimization' PSO. The term 'particle' means any natural agent that describes the `swarm' behavior. The PSO model is a particle simulation concept, and was first proposed by Eberhart and Kennedy (Eberhart, R.C. et al. 1995). Based upon a mathematical description of the social behavior of swarms, it has been shown that this algorithm can be efficiently generated to find good solutions to a certain number of complicated situations such as, for instance, the static optimization problems, the topological optimization and others (Parsopoulos, K.E. et al., 2001a); (Parsopoulos, K.E.et al. 2001b); (Fourie, P.C. et al., 2000); (Fourie, P.C. et al., 2001). Since then, several variants of the PSO have been developed (Eberhart,R.C. et al 1996); (Kennedy, J. et al., 1998); (Kennedy, J. et al., 2001); (Shi, Y.H. et al. 2001); (Shi, Y. et al. 1998a.); (Shi, Y.H. et al., 1998b); (Clerc, M. 1999). It has been shown that the question of convergence of the PSO algorithm is implicitly guaranteed if the parameters are adequately selected (Eberhart, R.C. et al.1998); (Cristian, T.I. 2003). Several kinds of problems solving start with computer simulations in order to find and analyze the solutions which do not exist analytically or specifically have been proven to be theoretically intractable.

The particle swarm treatment supposes a population of individuals designed as real valued vectors - particles, and some iterative sequences of their domain of adaptation must be established. It is assumed that these individuals have a social behavior, which implies that the ability of social conditions, for instance, the interaction with the neighborhood, is an important process in successfully finding good solutions to a given problem.

The strategy of the PSO algorithm is summarized as follows: We assume that each agent (particle) i can be represented in a N dimension space by its current position $x_i = \left(x_{i1}, x_{i2}, \ldots, x_{iN}\right)$ and its corresponding velocity. Also a memory of its personal (previous) best position is represented by, $p = \left(p_{i1}, p_{i2}, \ldots, p_{iN}\right)$ called (pbest), the subscript i range from 1 to s, where s indicates the size of the swarm. Commonly, each particle localizes its best value so far (pbest) and its position and consequently identifies its best value in the group (swarm), called also (sbest) among the set of values (pbest).

The velocity and position are updated as

$$v_{ij}^{k+1} = w_j v_{ij}^k + c_1 r_1^k \left[(pbest)_{ij}^k - x_{ij}^k\right] + c_2 r_2^k \left[(sbest)_{ij}^k - x_{ij}^k\right] \tag{7}$$

$$x_{ij}^{k+1} = v_{ij}^{k+1} + x_{ij}^k \tag{8}$$

where are the position and the velocity vector of particle i respectively at iteration k + 1, c_1 et c_2 are acceleration coefficients for each term exclusively situated in the range of 2--4, w_{ij} is the inertia weight with its value that ranges from 0.9 to 1.2, whereas r_1, r_2 are uniform random numbers between zero and one. For more details, the double subscript in the relations (7) and (8) means that the first subscript is for the particle i and the second one is for the dimension j. The role of a suitable choice of the inertia weight w_{ij} is important in the success of the PSO. In the general case, it can be initially set equal to its maximum value, and progressively we decrease it if the better solution is not reached. Too often, in the relation (7), w_{ij} is replaced by w_{ij} / σ, where σ denotes the constriction factor that controls the velocity of the particles. This algorithm is successively accomplished with the following steps (Zerarka, A. et al., 2006):

1. Set the values of the dimension space N and the size s of the swarm (s can be taken randomly).
2. Initialize the iteration number k (in the general case is set equal to zero).
3. Evaluate for each agent, the velocity vector using its memory and equation (7), where pbest and sbest can be modified.
4. Each agent must be updated by applying its velocity vector and its previous position using equation [8].
5. Repeat the above step (3, 4 and 5) until a convergence criterion is reached.

The practical part of using PSO procedure will be examined in the following section, where we'll interpolate Runge's "bell", with two manners; using Chebyshev interpolation approach and PSO approach, all while doing a comparison.

4.2 PSO distribution

So the problem is the choice of the points of interpolation so that quantity $\phi_n(x)$ deviates from zero on [a, b] the least possible.

Particle Swarm Optimization was used to find the global minimum of the maximum value of product $\phi_n(x)$, where very x is represented as a particle in the swarm.

The PSO parameter values that were used are given in Table 1.

Parameter	Setting
Population size	20
Number of iterations	500
C1 and C2	0.5
Inertial Weight	1.2 to 0.4
Desired Accuracy	10-5

Table 1. Particle Swarm Parameter Setting used in the present study

The best interpolating points x generated by PSO algorithm for polynomial of degree 5 and 10 respectively for example are:

Chebyshev	Points generated with PSO
-5.0000	-5.0000
-3.9355	-4.0451
-2.9041	-1.5451
0.9000	1.5451
3.9355	4.0451
5.0000	5.0000

Table 2 Polynomial of degree 5

Chebyshev	Points generated with PSO
-5.0000	-5.0000
-4.2900	-4.7553
-4.0251	-4.0451
-2.6500	-2.9389
-1.4000	-1.5451
0.0000	-0.0000
1.4000	1.5451
2.6500	2.9389
4.0451	4.0451
4.2900	4.7553
5.0000	5.0000

Table 3. Polynomial of degree 10

5. Comparison of interpolation methods

How big an effect can the selection of points have? Fig. 4 and Fig. 5 shows Runge's "bell" function interpolated over [-5, 5] using equidistant points, points selected from the Chebyshev distribution, and a new method called PSO. The polynomial interpolation using Chebyshev points does a much better job than the interpolation using equidistant points, but neither does as well as the PSO method.

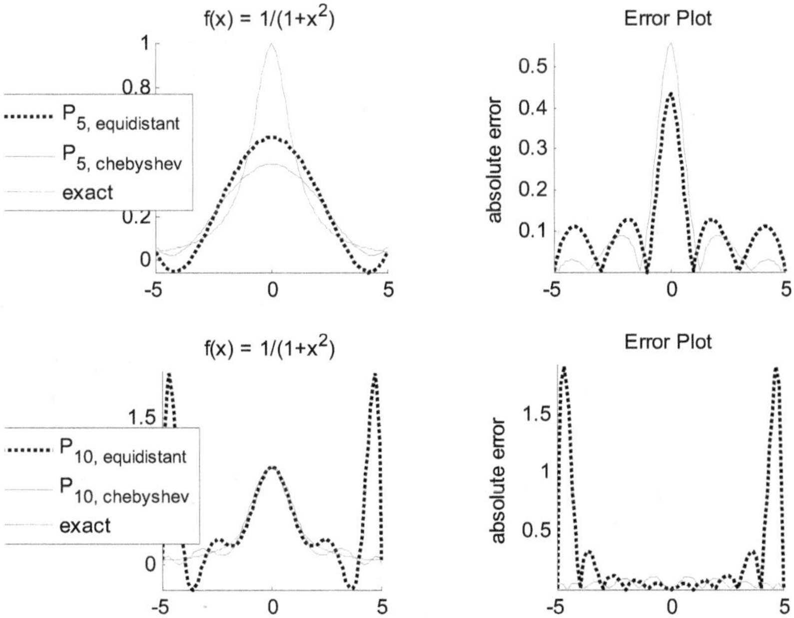

Figure 4. Comparison of interpolation polynomials for equidistant and Chebyshev sample points

Comparing Fig. 4, we see that the maximum deviation of the Chebyshev polynomial from the true function is considerably less than that of Lagrange polynomial with equidistant nodes. It can also be seen that increasing the number of the Chebyshev nodes—or, equivalently, increasing the degree of Chebyshev polynomial—makes a substantial contribution towards reducing the approximation error.
Comparing Fig. 5, we see that the maximum deviation of the PSO polynomial from the true function is considerably less than that of Chebyshev polynomial nodes. It can also be seen that increasing the number of the PSO nodes—or, equivalently, increasing the degree of PSO polynomial—makes a substantial contribution towards reducing the approximation error.

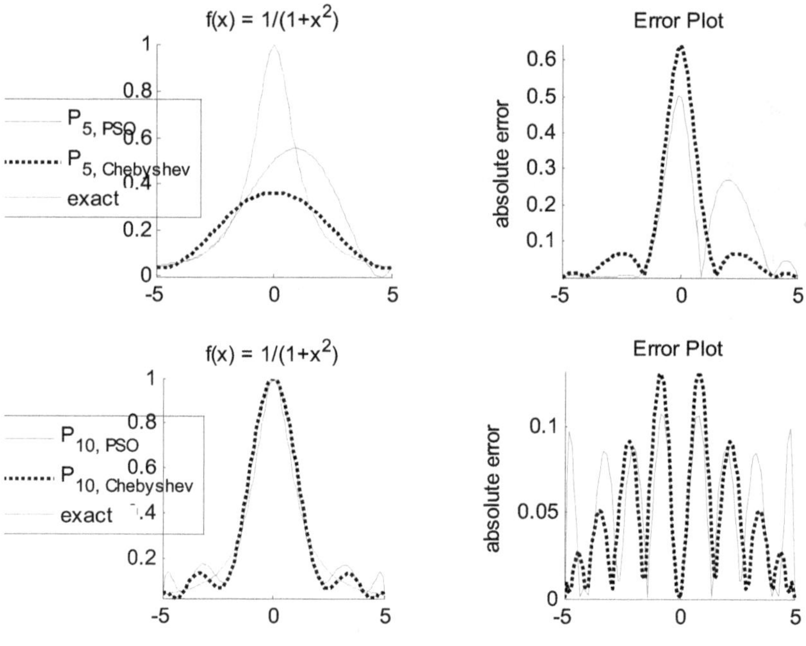

Figure 5. Comparison of interpolation polynomials for PSO and Chebyshev sample points

In this study we take as measure of the error of approximation the greatest vertical distance between the graph of the function and that of the interpolating polynomial over the entire interval under consideration (Fig. 4 and Fig. 5).
The calculation of error gives

Degree	Error points equidistant	Error points Chebychev	Error points PSO
5	0.4327	0.6386	0.5025
10	1.9156	0.1320	0.1076
15	2.0990	0.0993	0.0704
20	58.5855	0.0177	0.0131

Table 2. The error

6. Conclusion

The particle swarm optimization is used to investigate the best interpolating points. Some good results are obtained by using the specific PSO approach. It is now known that the PSO scheme is powerful, and easier to apply specially for this type of problems. Also, the PSO method can be used directly and in a straightforward manner. The performance of the scheme shows that the method is reliable and effective.

7. References

Clerc, M. (1999). The swarm and the queen: towards a deterministic and adaptive particle swarm optimization, *Proceedings of the 1999 IEEE Congress on Evolutionary Computation, pp.1951 – 1957,* Washington DC.

Cristian, T.I. (2003). The particle swarm optimization algorithm: convergence analysis and parameter selection, *Information Processing Letters,* Vol. 85, No. 6, pp.317--325.

David Kincaid and Ward Cheney, (2002). *Numerical Analysis: Mathematics of Scientific Computing.* Brooks/Cole.

Eberhart, R.C. and Kennedy, J. (1995). A new optimizer using particles swarm theory', *Sixth International Symposium on Micro Machine and Human Science,* pp.39--43, *Nagoya,* Japan.

Eberhart, R.C. and Shi, Y. (1998). Parameter selection in particle swarm optimization, in Porto, V.W.,

Eberhart, R.C. et al (1996). Computational Intelligence PC Tools, *Academic Press Professional,* Boston.

Fourie, P.C. and Groenwold, A.A. (2000). Particle swarms in size and shape optimization', *Proceedings of the International Workshop on Multi-disciplinary Design Optimization,* August 7--10, pp.97—106, Pretoria, South Africa.

Fourie, P.C. and Groenwold, A.A. (2001). Particle swarms in topology optimization', *Extended Abstracts of the Fourth World Congress of Structural and Multidisciplinary Optimization,* June 4--8, pp.52, 53, Dalian, China.

Hammer, R. Et al (1995). Numerical Toolbox for Verified Computing I, *Springer Verlag,* Berlin.

Kennedy, J. 1998. The behavior of particles, *Evol. Progr.,* Vol. VII, pp.581-587.

Kennedy J. and Eberhart, R.C, (1995). Particle swarm optimization, *Proc. IEEE Int. Conf. Neural Networks, Piscataway,* NJ, pp.1942--1948, USA.

Kennedy, J. and Eberhart, R.C. (2001). Swarm Intelligence, *Morgan Kaufmann Publishers,* San Francisco.

Kennedy, J. and Spears, W.M. (1998). Matching algorithms to problems: an experimental test of the particle swarm and some genetic algorithms on the multimodal problem generator, *Proceedings of the (1998) IEEE International Conference on Evolutionary Computation,* Anchorage, May 4--9, Alaska.

Kulisch, U. and Miranker, W.L. (1983). A New Approach to Scientific Computation, Academic Press, New York.

L. N. Trefethen. Spectral Methods in Matlab. *SIAM,* (2000). 9, Philadelphia

L. Djerou, M. Batouche, N. Khelil and A.Zerarka, *(2007).* Towards the best points of interpolation using Particles swarm optimization approach, *in proceedings of IEEE Congress of Evolutionary Computing, CEC 2007, pp. 3211-3214, Singapore.*

Maron, M. and Lopez, R. (1991). Numerical Analysis, *Wadsworth Publishing Company*, Belmont, California.

Parsopoulos, K.E. and Vrahatis, M.N. (2001). `Modification of the particle swarm optimizer for locating all the global minima', in Kurkova, V. et al. (Eds.): *Artificial Neural Networks and Genetic Algorithms, Springer*, pp.324--327, New York.

Parsopoulos, K.E. et al, (2001a). Objective function stretching to alleviate convergence to local minima, *Nonlinear Analysis TMA*, Vol. 47, pp.3419--3424.

Parsopoulos, K.E.et al (2001b). Stretching technique for obtaining global minimizers through particle swarm optimization, *Proceedings of the PSO Workshop*, pp.22--29, Indianapolis, USA.

Roland E. Larson, Robert P. Hostetler, Bruch H. Edwards and David E. Heyd, (1994)., *Calculus with Analytic Geometry*. D. C. Heath and Company.

Saravanan, N., Waagen, D. and Eiben, A.E. (Eds.): *Lecture Notes in Computer Science-Evolutionary Programming VII, Springer*, Vol. 1447, pp.591--600.

Shi, Y. and Eberhart, R.C. (1998a). A modified particle swarm optimizer, *Proceedings of the 1998 IEEE International Conference on Evolutionary Computation*, May 4—9, Anchorage, Alaska.

Shi, Y.H. and Eberhart, R.C. (1998b). Parameter selection in particle swarm optimization, Evolutionary Programming VII, *Lecture Notes in Computer Science*, pp.591--600.

Shi, Y.H. and Eberhart, R.C. (2001). Fuzzy adaptive particle swarm optimization', *IEEE Int. Conf. on Evolutionary Computation*, pp.101--106.

Zerarka, A., Khelil, N. (2006). A generalized integral quadratic method: improvement of the solution for one dimensional Volterra integral equation using particle swarm optimization, *Int. J. Simulation and Process Modeling*, Vol. 2, Nos. 1/2, pp.152-163.

2

Particle Swarm Optimization with External Archives for Interactive Fuzzy Multiobjective Nonlinear Programming

Takeshi Matsui, Masatoshi Sakawa, Kosuke Kato and Koichi Tamada
Hiroshima University
Japan

1. Introduction

In general nonlinear programming problems to find a solution which minimizes an objective function under given constraints, one whose objective function and constraint region are convex is called a convex programming problem. For such convex programming problems, there have been proposed many efficient solution method as the successive quadratic programming method and the general gradient method. Unfortunately, there have not been proposed any decisive solution method for nonconvex programming problems. As practical solution methods, meta-heuristic optimization methods as the simulated annealing method and the genetic algorithm have been proposed.

In recent years, however, more speedy and more accurate optimization methods have been desired because the size of actual problems has been increasing.

As a new optimization method, particle swarm optimization (PSO) was proposed (Kennedy & Eberhart, 1995). PSO is a search method simulating the social behavior that each individual in the population acts by using both the knowledge owned by it and that owned by the population, and they search better points by constituting the population. The authors proposed a revised PSO (rPSO) by incorporating the homomorphous mapping and the multiple stretching technique in order to deal with shortcomings of the original PSO as the concentration to local solution and the inapplicability of constrained problems (Matsui et al., 2008).

In recent years, with the diversification of social requirements, the demand for the programs with multiple objective functions, which may be conflicting with each other, rather than a single-objective function, has been increasing (e.g. maximizing the total profit and minimizing the amount of pollution in a production planning). Since there does not always exist a complete optimal solution which optimizes all objectives simultaneously for multiobjective programming problems, the Pareto optimal solution or non-inferior solution, is defined, where a solution is Pareto optimal if any improvement of one objective function can be achieved only at the expense of at least one of the other objective functions. For such multiobjective optimization problems, fuzzy programming approaches (e.g. (Zimmermann, 1983), (Rommelfanger, 1996), considering the imprecise nature of the DM's judgments in multiobjective optimization problems, seem to be very applicable and promising. In the application of the fuzzy set theory into multiobjective linear programming problems started

(Zimmermann, 1978), it has been implicitly assumed that the fuzzy decision or the minimum-operator of (Bellman & Zadeh ,1970) is the proper representation of the DM's fuzzy preferences. Thereby, M. Sakawa et al. have proposed interactive fuzzy satisficing methods to derive satisficing solutions for the decision maker along with checking the local preference of the decision maker through interactions for various multiobjective programming problems (Sakawa et al, 2002).

In this paper, focusing on multiobjective nonlinear programming problems, we attempt to derive satisficing solutions through the interactive fuzzy satisficing method. Since problems solved in the interactive fuzzy satificing method for multiobjective nonlinear programming problems are nonlinear programming problems, we adopt rPSO (Matsui et al, 2008) as solution methods to them. In particular, we consider measures to improve the performance of rPSO in applying it to solving the augmented minimax problem.

2. Multiobjective nonlinear programming problems

In this paper, we consider multiobjective nonlinear programming problem as follows:

$$\begin{array}{ll} \text{minimize} & f_l(\boldsymbol{x}),\ l=1,2,\ldots,k \\ \text{subject to} & g_i(\boldsymbol{x}) \leq 0,\ i=1,2,\ldots,m \\ & l_j \leq x_j \leq u_j,\ j=1,2,\ldots,n \\ & \boldsymbol{x}=(x_1,x_2,\ldots,x_n)^T \in R^n \end{array} \tag{1}$$

where $f_l(\cdot)$, $g_i(\cdot)$ are linear or nonlinear functions, l_j and u_j are the lower limit and the upper limit of each decision variable x_j. In addition, we denote the feasible region of (1) by X.

3. An interactive fuzzy satisficing method

In order to consider the imprecise nature of the decision maker's judgments for each objective function in (1), if we introduce the fuzzy goals such as ``$f_l(\boldsymbol{x})$ should be substantially less than or equal to a certain value'', (1) can be rewritten as:

$$\underset{\boldsymbol{x}\in X}{\text{maximize}} \quad (\mu_1(f_1(\boldsymbol{x})),\ldots,\mu_k(f_k(\boldsymbol{x}))) \tag{2}$$

where $\mu_l(\cdot)$ is the membership function to quantify the fuzzy goal for the l th objective function in (1).

Since (2) is regarded as a fuzzy multiobjective decision making problem, there rarely exists a complete optimal solution that simultaneously optimizes all objective functions. As a reasonable solution concept for the fuzzy multiobjective decision making problem, M. Sakawa defined M-Pareto optimality on the basis of membership function values by directly extending the Pareto optimality in the ordinary multiobjective programming problem (Sakawa, 1993). In the interactive fuzzy satisficing method, in order to generate a candidate for the satisficing solution which is also M-Pareto optimal, the decision maker is asked to specify the aspiration levels of achievement for all membership functions, called the reference membership levels (Sakawa, 1993). For the decision maker's reference membership levels $\bar{\mu}_l$, l=1, ..., k, the corresponding M-Pareto optimal solution, which is nearest to the requirements in the minimax sense or better than it if the reference membership levels are attainable, is obtained by solving the following augmented minimax problem (3).

$$\underset{x \in X}{\text{minimize}} \quad \max_{l=1,\ldots,k} \left\{ (\bar{\mu}_l - \mu_l(f_l(\boldsymbol{x}))) + \rho \sum_{i=1}^{k} (\bar{\mu}_i - \mu_i(f_i(\boldsymbol{x}))) \right\} \tag{3}$$

where ρ is a sufficiently small positive number.
We can now construct the interactive algorithm in order to derive the satisficing solution for the decision maker from the M-Pareto optimal solution set. The procedure of an interactive fuzzy satisficing method is summarized as follows.

Step 1:
Under a given constraint, minimal value and maximum one of each objective function are calculated by solving following problems.

$$\underset{x \in X}{\text{minimize}} \quad f_l(\boldsymbol{x}),\ l=1, 2, \ldots, k \tag{4}$$

$$\underset{x \in X}{\text{maximize}} \quad f_l(\boldsymbol{x}),\ l=1, 2, \ldots, k \tag{5}$$

Step 2:
In consideration of individual minimal value and maximum one of each objective function, the decision maker subjectively specifies membership functions $\mu_l(f_l(\boldsymbol{x}))$, l=1, …, k to quantify fuzzy goals for objective functions. Next, the decision maker sets initial reference membership function values $\bar{\mu}_l$, l=1, …, k.

Step 3:
We solve the following augmented minimax problem corresponding to current reference membership function values (3).

Step 4:
If the decision maker is satisfied with the solution obtained in Step 3, the interactive procedure is finished. Otherwise, the decision maker updates reference membership function values $\bar{\mu}_l$, l=1, 2, …, k based on current membership function values and objective function values, and return to Step 3.

4. Particle swarm optimization

Particle swarm optimization (Kennedy & Eberhart, 1995) is based on the social behavior that a population of individuals adapts to its environment by returning to promising regions that were previously discovered (Kennedy & Spears, 1998). This adaptation to the environment is a stochastic process that depends on both the memory of each individual, called particle, and the knowledge gained by the population, called swarm. In the numerical implementation of this simplified social model, each particle has four attributes: the position vector in the search space, the velocity vector and the best position in its track and the best position of the swarm. The process can be outlined as follows.

Step 1:

Generate the initial swarm involving N particles at random.

Step 2:

Calculate the new velocity vector of each particle, based on its attributes.

Step 3:

Calculate the new position of each particle from the current positon and its new velocity vector.

Step 4:

If the termination condition is satisfied, stop. Otherwise, go to Step 2.

To be more specific, the new velocity vector of the i-th particle at time t, $\boldsymbol{v}_i^{t+1}$ is calculated by the following scheme introduced by (Shi & Eberhart, 1998).

$$\boldsymbol{v}_i^{t+1} := \omega^t \boldsymbol{v}_i^t + c_1 R_1^t (\boldsymbol{p}_i^t - \boldsymbol{x}_i^t) + c_2 R_2^t (\boldsymbol{p}_g^t - \boldsymbol{x}_i^t) \tag{6}$$

In (6), R_1^t and R_2^t are random numbers between 0 and 1, p_i^t is the best position of the i-th particle in its track and p_g^t is the best position of the swarm. There are three problem dependent parameters, the inertia of the particle ω^t, and two trust parameters c_1, c_2. Then, the new position of the i-th particle at time t, x_i^{t+1}, is calculated from (7).

$$x_i^{t+1} := x_i^t + v_i^{t+1} \tag{7}$$

where x_i^t is the current position of the i-th particle at time t. The i-th particle calculates the next search direction vector $\boldsymbol{v}_i^{t+1}$ by (6) in consideration of the current search direction vector $\boldsymbol{v}_i^t$, the direction vector going from the current search position x_i^t to the best position in its track p_i^t and the direction vector going from the current search position x_i^t to the best position of the swarm p_g^t, moves from the current position x_i^t to the next search position x_i^{t+1} calculated by (7). The parameter ω^t controls the amount of the move to search globally in early stage and to search localy by decreasing ω^t gradually.

The searching procedure of PSO is shown in Fig. 1.

Comparing the evaluation value of a particle after movement, $f(\boldsymbol{x}_i^{t+1})$, with that of the best position in its track, $f(p_i^t)$, if $f(\boldsymbol{x}_i^{t+1})$ is better than $f(p_i^t)$, then the best position in its track is updated as $p_i^t := x_i^{t+1}$. Futhermore, if $f(p_i^{t+1})$ is better than $f(p_g^t)$, then the best position in the swarm is updated as $p_g^{t+1} := p_i^{t+1}$.

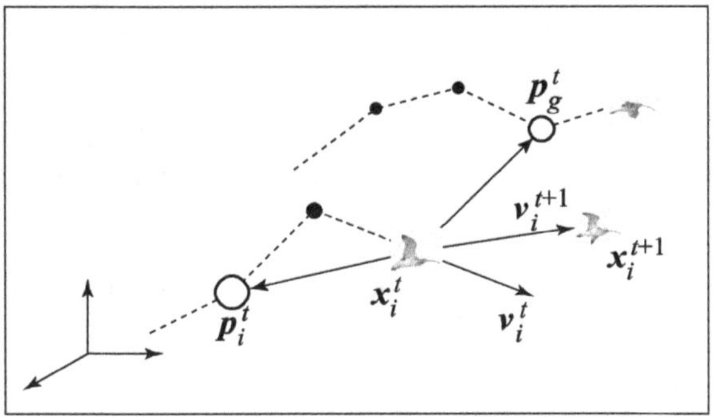

Figure 1. Movement of a particle in PSO

In the original PSO method, however, there are drawbacks that it is not directly applicable to constrained problems and it is liable to stopping around local optimal solutions.

To deal with these drawbacks of the original PSO method, we incorporate the bisection method and a homomorphous mapping to carry out the search considering constraints.

In addition, we proposed the multiple stretching technique and modified move schemes of particles to restrain the stopping around local optimal solutions (Matsui et al., 2008).

Thus, we applied rPSO for interactive fuzzy multiobjective nonlinear programming problems and proposed multiobjective revised PSO (MOrPSO) method incorporating move scheme to the nondominated particle in order to search effectively for the augmented minimax problmes (Matsui et al., 2007). In the application of large-scale augmented minimax problem, MOrPSO method is superior than rPSO method on efficiency. On the other hand, MOrPSO method is inferior on accuracy.

5. Improvement of MOrPSO

We show the results of the applicaltion of the original rPSO (Matsui et al., 2008) and MOrPSO (Matsui et al., 2007) to the augmented minimax problem for multiobjective nonlinear programming problem with l = 2, n = 55 and m = 100 in Table 1. In these experiments we set the swarm size N = 70, the maximal search generation number $T_{\max}$ = 5000. In addition, we use the following membership functions: $\bar{\mu}_1 = 1.0$, $\bar{\mu}_2 = 1.0$.

method	objective function value (minimize)			computational time (sec)
	best	average	worst	
rPSO	0.3464	0.4471	0.5632	144.45
MOrPSO	0.3614	0.4095	0.4526	129.17

Table 1. Results of the application to the augmented minimax problem

From Table 1, MOrPSO method is superior than rPSO method on efficiency in the average value, the worst one and computational time. However, the best value of MOrPSO method is worse than that of rPSO method, MOrPSO method is inferior on accuracy. We consider

the case that the search accuracy turns worse incorporating the direction to nondominated particle (approximate M-Pareto optimal solution) in MOrPSO method.
In this paper, we improve the search accuracy incorporating external archives to record nondominated particles in the swarm. Here, as recorded nondominated particle increases in archives, computational time increases in order to judge whether a particle is nondominated.
Therefore, there is many computational time that we record all the nondominated particles to archives. Thus we divide membership function space with hypercube shown in Fig. 2 and record a number of nondominated particle included in each hypercube.

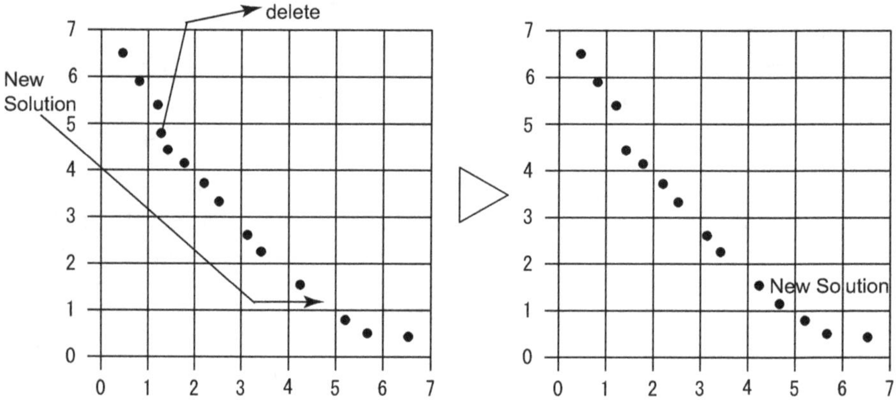

Figure 2. reduction of archives with grid (l=2)

When a number of nondominated particle recorded in archives is greater than a fixed number, we delete one particle from hypercube with many numbers of nondominated particle and record new solution (particle). We consider that can reduce computational time and express approximate M-Pareto optimal front by a few particles incorporating reduction of archives. We show the results of the application of MOrPSO method incorporating reduction of archives (MOrPSO-1) to the above same problem in Table 2.

method	objective function value (minimize)			computational time (sec)
	best	average	worst	
MOrPSO-1	0.4379	0.4708	0.5030	166.78

Table 2. Results of the application to the augmented minimax problem

From Table 2, it is clear that all of the best, average, worst value and computational time obtained by MOrPSO-1 are worse than those obtained by MOrPSO. We consider that a particle moves using nondominated particle which is not useful since all nondominated particles in archives are used in search. Thus, in the membership function space, we consider that information of a particle exsiting far from the reference membership value is hard to contribute to search and introduce threshold value for selection of nondominated particle as shown in Fig. 3.

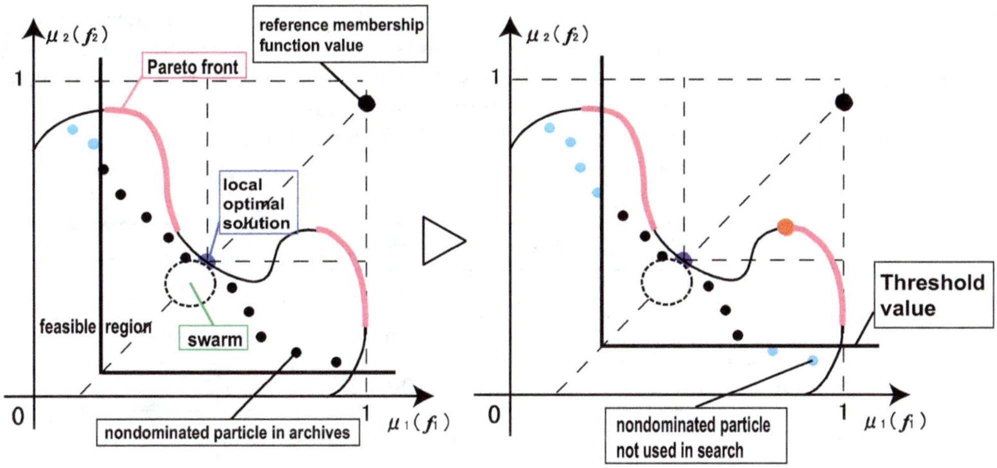

Figure 3. Limit of nondominated particle by threshold value (l=2)

We show the results of the application of MOrPSO method incorporating limitation by threshold value (MOrPSO-2) to the above same problem in Table 3.

method	objective function value (minimize)			computational time (sec)
	best	average	worst	
MOrPSO-2	0.2993	0.3430	0.3777	165.71

Table 3. Results of the application to the augmented minimax problem

From Table 3, in the application of MOrPSO method incorporating limitation by threshold value (MOrPSO-2), we can get better solutions in the sense of best, average and worst than those obtained by rPSO and MOrPSO.

In order to show the efficiency of the proposed PSO, we consider the multiobjective nonlinear programming problem with l = 2 and n = 100. In these experiments, we set the swarm size N = 100, the maximal search generation number $T_{\max}$ = 5000. In addition, we use the following reference membership function values: $\bar{\mu}_1 = 1.0$, $\bar{\mu}_2 = 1.0$.

method	objective function value (minimize)			computational time (sec)
	best	average	worst	
rPSO [6]	0.2547	0.2783	0.3251	26.91
MOrPSO [7]	0.1950	0.2033	0.2208	32.58
MOrPSO-2	0.2018	0.2320	0.2711	28.11

Table 4. Results of the application to the augmented minimax problem

From Table 4, it is clear that all of the best, average, worst value and computational time obtained by MOrPSO-2 are worse than those obtained by MOrPSO. We consider that it occurs to make no use of the information of nondominiated particle in search since there is a few nondominiated particle information stored in archives of MOrPSO-2 and only the best value of each objective function and the best value limb of the augmented minimax problem are saved. Therefore, we propose MOrPSO with external archives (MOrPSO-EA) using nondominated particle in the swarm same as MOrPSO in order to store various nondominated particle as possible in normal search. And the results of the application are shown in Table 5.

method	objective function value (minimize)			computational time (sec)
	best	average	worst	
MOrPSO	0.1950	0.2033	0.2208	32.58
MOrPSO-EA	0.1746	0.1787	0.1842	29.01

Table 5. Results of the application to the augmented minimax problem

From Table 5, in the application of MOrPSO-EA, we can get better solutions in the sense of best, average and worst than those obtained by MOrPSO.

6. Numerical examples

In MOrPSO, it searches globaly in the early generation and localy decreasing ω^t. However, we consider that necessity to search globaly in the early generation is low after the second time since the information of nondominated particle to current generation is stored in archives in proposed MOrPSO. Therefore, we consider that the proposed MOrPSO can search localy in the early generation.

In order to show the efficiency of the proposed MOrPSO, we consider the multiobjective nonlinear programming problem with l = 2 and n = 100 and m = 55. In these experiments, we set the maximal search generation number of MOrPSO and the proposed MOrPSO (MOrPSO-EA) in the 1st interactive $T_{\max}$ = 5000 and in the 2nd and 3rd interactive $T_{\max}$ = 3000. We show the results of the application are shown in Table 6 and 7.

interactive	1st	2nd	3rd
$\bar{\mu}_1$	1.0	1.0	0.85
$\bar{\mu}_2$	1.0	0.7	0.7
$\mu_1(x)$	0.6157	0.8003	0.7575
$\mu_2(x)$	0.6157	0.5003	0.6075
minimax value	0.3844	0.1997	0.0925
time (sec)	127.20	128.89	131.25

Table 6. Interactive fuzzy programming through MOrPSO

interactive	1st	2nd	3rd
$\bar{\mu}_1$	1.0	1.0	0.85
$\bar{\mu}_2$	1.0	0.7	0.7
$\mu_1(x)$	0.7183	0.8458	0.8042
$\mu_2(x)$	0.7183	0.5458	0.6542
minimax value	0.2817	0.1542	0.0458
time (sec)	157.37	98.58	101.16

Table 7. Interactive fuzzy programming through MOrPSO-EA (proposed)

From Table 6 and 7, MOrPSO-EA is superior than MOrPSO on accuracy. In addition, we can decrease total computational time by reducing the maximal search generation number.

6. Conclusion

In this research, we focused on multiobjective nonlinear programming problems and proposed a new MOrPSO technique which is accuracy for in applying the interactive fuzzy satisficing method. In particular, considering the features of augmented minimax problems solved in the interactive fuzzy satisficing method, we incorporated use of external archives, reduction of archives and the limitation of threshold value. Finally, we showed the efficiency of the proposed MOrPSO by applying it to numerical examples.

7. References

R.E. Bellman & L.A. Zadeh. (1970). Decision making in a fuzzy environment, *Management Science*, Vol. 17, pp. 141-164

C.A.C Coello, G.T. Pulido, M.S. Lechuga. (2004). Handling multiple objectives with particle swarm optimization, *IEEE Transactions on Evolutionary Computation*, Vol. 8, No. 3, pp. 256-279

V.L. Huang, A.K. Qin, K. Deb, E. Zitzler, P.N. Suganthan, J.J. Liang, M. Preuss, S. Huband. (2007). Problem definitions for performance assessment on multi-objective optimization algorithms, *http://www.ntu.edu. sg/home/EPNSugan/index-files/CEC-07/CEC-07-TR-13-Feb.pdf*

J. Kennedy & R.C. Eberhart. (1995). Particle Swarm Optimization, *Proceedings of IEEE International Conference on Neural Networks*, pp. 1942-1948

T. Matsui, K. Kato, M. Sakawa, T. Uno, K. Morihara. (2008). Nonlinear Programming Based on Particle Swarm Optimization, In: *Advances in Industrial Engineering and Operations Research*, Alan H. S. Chan, Sio-Iong Ao (Ed.), pp. 173-183, Springer, New York, 2008.

T. Matsui, M. Sakawa, K. Kato, T. Uno, K. Tamada. (2007). An interactive fuzzy satisficing method through particle swarm optimization for multiobjective nonlinear programming problems, *Proceedings of the 2007 IEEE Symposium Series on Computational Intelligence*, pp. 71-76

H. Rommelfanger. (1996). Fuzzy linear programming and applications, *European Journal of Operational Research*, vol. 92, pp. 512-528

M. Sakawa. (1993). *Fuzzy Sets and Interactive Multiobjective Optimization*, Plenum Press, New York

M. Sakawa, K. Kato, T. Suzuki. (2002). An interactvie fuzzy satisficing method for multiobjective non-convex programming problems through genetic algorithms, *Proceedings of 8th Japan Society for Fuzzy Theory and Systems Chugoku / Shikoku Branch Office Meeting*, pp. 33-36

Y.H. Shi & R.C. Eberhart. (1998). A modified particle swarm optimizer, *Proceedings of IEEE International Conference on Evolutionary Computation*, pp. 69-73

H.-J. Zimmermann. (1978). Fuzzy programming and linear programming with several objective functions, *Fuzzy Sets and Systems*, Vol. 1 pp. 45-55

H.-J. Zimmermann. (1983). Fuzzy mathematical programming, *Computers & Operations Research*, Vol. 10, pp. 291-298

3

Finding Base-Station Locations in Two-Tiered Wireless Sensor Networks by Particle Swarm Optimization*

Tzung-Pei Hong[1, 2], Guo-Neng Shiu[3] and Yeong-Chyi Lee[4]
[1]*Department of Computer Science and Information Engineering, National University of Kaohsiung*
[2]*Department of Computer Science and Engineering, National Sun Yat-sen University*
[3]*Department of Electrical Engineering, National University of Kaohsiung*
[4]*Department of Information Management, Cheng-Shiu University*
Taiwan

1. Abstract

In wireless sensor networks, minimizing power consumption to prolong network lifetime is very crucial. In the past, Pan *et al.* proposed two algorithms to find the optimal locations of base stations in two-tiered wireless sensor networks. Their approaches assumed the initial energy and the energy-consumption parameters were the same for all application nodes. If any of the above parameters were not the same, their approaches could not work. Recently, the PSO technique has been widely used in finding nearly optimal solutions for optimization problems. In this paper, an algorithm based on particle swarm optimization (PSO) is thus proposed for general power-consumption constraints. The proposed approach can search for nearly optimal BS locations in heterogeneous sensor networks, where application nodes may own different data transmission rates, initial energies and parameter values. Experimental results also show the good performance of the proposed PSO approach and the effects of the parameters on the results. The proposed algorithm can thus help find good BS locations to reduce power consumption and maximize network lifetime in two-tiered wireless sensor networks.
Keywords: wireless sensor network, network lifetime, energy consumption, particle swarm optimization, base station.

2. Introduction

Recently, a two-tiered architecture of wireless sensor networks has been proposed and become popular [1]. It is motivated by the latest advances in distributed signal processing

---

* This is a modified and expanded version of the paper "A PSO heuristic algorithm for base-station locations," presented at The Joint Conference of the Third International Conference on Soft Computing and Intelligent Systems and the Seventh International Symposium on Advanced Intelligent Systems, 2006, Japan.

and source coding and can offer a more flexible balance among reliability, redundancy and scalability of wireless sensor networks. A two-tiered wireless sensor network, as shown in Figure 1, consists of sensor nodes (SNs), application nodes (ANs), and one or several base stations (BSs).

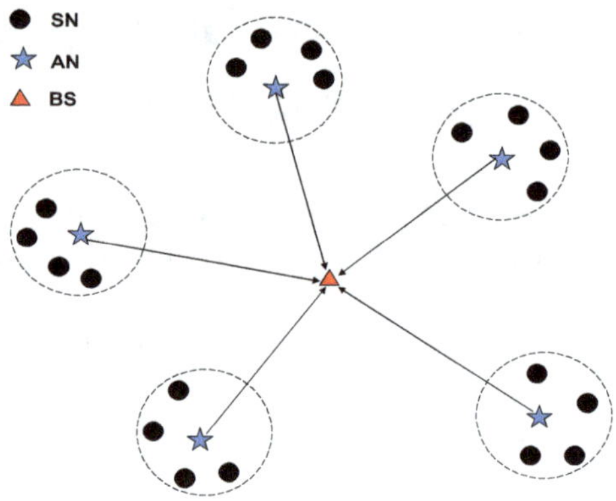

Figure 1. A two-tiered architecture of wireless sensor networks

Sensor nodes are usually small, low-cost and disposable, and do not communicate with other sensor nodes. They are usually deployed in clusters around interesting areas. Each cluster of sensor nodes is allocated with at least one application node. Application nodes possess longer-range transmission, higher-speed computation, and more energy than sensor nodes. The raw data obtained from sensor nodes are first transmitted to their corresponding application nodes. After receiving the raw data from all its sensor nodes, an application node conducts data fusion within each cluster. It then transmits the aggregated data directly to the base station or via multi-hop communication. The base station is usually assumed to have unlimited energy and powerful processing capability. It also serves as a gateway for wireless sensor networks to exchange data and information to other networks. Wireless sensor networks usually have some assumptions for SNs and ANs. For instance, each AN may be aware of its own location through receiving GPS signals [11] and its own energy.

In the past, many approaches were proposed to efficiently utilize energy in wireless networks. For example, appropriate transmission ways were designed to save energy for multi-hop communication in ad-hoc networks [16][10][5][19][7][6][20]. Good algorithms for allocation of base stations and sensors nodes were also proposed to reduce power consumption [12][15][16][8][9]. Thus, a fundamental problem in wireless sensor networks is to maximize the system lifetime under some given constraints. Pan *et al.* proposed two algorithms to find the optimal locations of base stations in two-tiered wireless sensor networks [13]. Their approaches assumed the initial energy and the energy-consumption parameters were the same for all ANs. If any of the above parameters were not the same, their approaches could not work.

In this paper, an algorithm based on particle swarm optimization (PSO) is proposed to find the base-station locations for general power-consumption constraints. The PSO technique was proposed by Eberhart and Kennedy in 1995 [2][3] and has been widely used in finding solutions for optimization problems. Some related researches about its improvement and applications has also been proposed [4][14][17][18]. It maintains several particles (each represents a solution) and all the particles continuously move in the search space according to their own local optima and the up-to-date global optimum. After a lot of generations, the optimal solution or an approximate optimal solution is expected to be found. The proposed approach here can search for nearly optimal BS locations in heterogeneous sensor networks. Experimental results also show the performance of the proposed PSO approach on finding the BS locations and the effects of the parameters on the results.

The remaining parts of this paper are organized as follows. Some related works about finding the locations of base stations in a two-tiered wireless networks is reviewed in Section 3. An algorithm based on PSO to discover base stations in a two-tiered wireless networks is proposed in Section 4. An example to illustrate the proposed algorithm is given in Section 5. Experimental results for demonstrating the performance of the algorithm and the effects of the parameters are described in Section 6. Conclusions are stated in Section 7.

3. Review of Related Works

As mentioned above, a fundamental problem in wireless sensor networks is to maximize the system lifetime under some given constraints. Pan *et al.* proposed two algorithms to find the optimal locations of base stations in two-tiered wireless sensor networks [13]. The first algorithm was used to find the optimal locations of base stations for homogenous ANs, and the second one was used for heterogeneous ANs. Homogenous ANs had the same data transmission rate and heterogeneous ANs might have different data transmission rates. In their paper, only the energy in ANs was considered. If a single SN ran out of energy, its corresponding AN might still have the capability to collect enough information. However, if an AN ran out of energy, the information in its coverage range would be completely lost, which was dangerous to the whole system.

Let d be the Euclidean distance from an AN to a BS, and r be the data transmission rate. Pan *et al.* adopted the following formula to calculate the energy consumption per unit time:

$$p(r,d) = r(\alpha_1 + \alpha_2 d^b)\,, \tag{1}$$

where a_1 is a distance-independent parameter, a_2 is a distance-dependent parameter, and b is the Euclidean dimension. The energy consumption thus relates to Euclidean distances and data transmission rates.

Pan *et al.* assumed each AN had the same a_1, a_2 and initial energy. For homogenous ANs, they showed that the center of the minimal circle covering all the ANs was the optimal BS location (with the maximum lifetime).

4. A General Base-Station Allocation Algorithm Based on PSO

The ANs produced by different manufacturers may own different data transmission rates, initial energies and parameter values. When different kinds of ANs exist in a wireless network, it is hard to find the optimal BS location. In this section, a heuristic algorithm based on PSO to search for optimal BS locations under general constraints is proposed. An initial set of particles is first randomly generated, with each particle representing a possible BS location. Each particle is also allocated an initial velocity for changing its state. Let $e_j(0)$ be the initial energy, r_j be the data transmission rate, a_{j1} be the distance-independent parameter, and a_{j2} be the distance-dependent parameter of the j-th AN. The lifetime l_{ij} of an application node AN_j for the i-th particle is calculated by the following formula:

$$l_{ij} = e_j(0)/r_j(\alpha_{j1} + \alpha_{j2}d_{ij}^b), \tag{2}$$

where d_{ij}^b is the b-order Euclidian distance from the j-th AN to the i-th particle. The fitness function used for evaluating each particle is thus shown below:

$$fitness(i) = \underset{j=1}{\overset{m}{Min}}\, l_{ij}, \tag{3}$$

where m is number of ANs. That is, each particle takes the minimal lifetime of all ANs as its fitness value. A larger fitness value denotes a longer lifetime of the whole system, meaning the corresponding BS location is better. The fitness value of each particle is then compared with that of its corresponding *pBest*. If the fitness value of the i-th particle is larger than that of $pBest_i$, $pBest_i$ is replaced with the i-th particle. The best $pBest_i$ among all the particles is chosen as the *gBest*. Besides, each particle has a velocity, which is used to change the current position. All particles thus continuously move in the search space. When the termination conditions are achieved, the final *gBest* will be output as the location of the base station. The proposed algorithm is stated below.

The proposed PSO algorithm for finding the best BS location:

- Input: A set of ANs, each AN_j with its location (x_j, y_j), data transmission rate r_j, initial energy $e_j(0)$, parameters a_{j1} and a_{j2}.
- Output: A BS location that will cause a nearly maximal lifetime in the whole system.
- Step 1: Initialize the fitness values of all *pBests* and the *gBest* to zero.
- Step 2: Randomly generate a group of n particles, each representing a possible BS location. Locations may be two-dimensional or three-dimensional, depending on the problems to be solved.
- Step 3: Randomly generate an initial velocity for each particle.
- Step 4: Calculate the lifetime l_{ij} of the j-th AN for the i-th particle by the following formula:

  $$l_{ij} = e_j(0)/r_j(\alpha_{j1} + \alpha_{j2}d_{ij}^b),$$

  where $e_j(0)$ is the initial energy, r_j is the data transmission rate, a_{j1} is a distance-independent parameter, a_{j2} is a distance-dependent parameter of the j-th AN, and d_{ij}^b is the b-order Euclidean distance from the i-th particle (BS) to the j-th AN.
- Step 5: Calculate the lifetime of the whole sensor system for the i-th particle as its fitness value ($fitness_i$) by the following formula:

  $$fitness(i) = \underset{j=1}{\overset{m}{Min}}\, l_{ij},$$

  where m is number of ANs and $i = 1$ to n.
- Step 6: Set $pBest_i$ as the current i-th particle if the value of $fitness(i)$ is larger than the current fitness value of $pBest_i$.
- Step 7: Set *gBest* as the best *pBest* among all the particles. That is, let:

$$\text{fitness of } pBest_k = max_{i=1}^{n} \text{ fitness of } pBest_i,$$

and set $gBest$=$pBest_k$.

- Step 8: Update the velocity of the i-th particle as:

$$V_{id}^{new} = w \times V_{id}^{old} + c_1 \times Rand_1() \times (pBest_{id} - x_{id}) + c_2 \times Rand_2() \times (gBest_d - x_{id})$$

  where x_{id}^{new} is the new velocity of the i-th particle at the d-th dimension, x_{id}^{old} is the current velocity of the i-th particle at the d-th dimension, w is the inertial weight, c_1 is the acceleration constant for particles moving to $pBest$, c_2 is the acceleration constant for particles moving to $gBest$, $Rand_1()$ and $Rand_2()$ are two random numbers among 0 to 1, x_{id} is the current position of the i-th particle at the d-th dimension, $pBest_{id}$ is the value of $pBest_i$ at the d-th dimension, and $gBest_d$ is the value of $gBest$ at the d-th dimension.
- Step 9: Update the position of the i-th particle as:

$$x_{id}^{new} = x_{id}^{old} + V_{id}^{new},$$

  where x_{id}^{new} and x_{id}^{old} are respectively the new position and the current position of the i-th particle at the d-th dimension.
- Step 10: Repeat Steps 4 to 9 until the termination conditions are satisfied.

In Step 10, the termination conditions may be predefined execution time, a fixed number of generation or when the particles have converged to a certain threshold.

5. An Example

In this section, a simple example in a two-dimensional space is given to explain how the PSO approach can be used to find the best BS location that will generate the nearly maximal lifetime in the whole system. Assume there are totally four ANs in this example and their initial parameters are shown in Table 1, where "Location" represents the two-dimensional coordinate position of an AN, "Rate" represents the data transmission rate, and "Power" represents the initially allocated energy. All a_{j1}'s are set at 0 and all a_{j2}'s at 1 for simplicity.

AN	Location	Rate	Power
1	(1, 10)	5	10000
2	(11, 0)	5	10000
3	(8, 7)	4	6400
4	(4, 3)	4	6400

Table 1. The initial parameters of ANs in the example

For the example, the proposed PSO algorithm proceeds as follows.

- Step 1: The initial fitness values of all $pBests$ and the $gBest$ are set to zero.
- Step 2: A group of n particles are generated at random. Assume n is set at 3 in this example for simplicity. Also assume the three initial particles randomly generated are located at (4, 7), (9, 5) and (6, 4). Figure 2 shows the positions of the given ANs and the initial particles, where the triangles represent the particles and the circles represent the ANs.

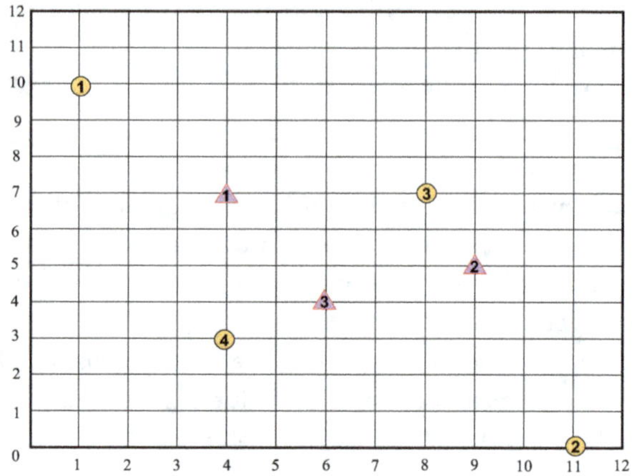

Figure 2. The positions of the given ANs and the initial particles

- Step 3: An initial velocity is randomly generated for each particle. In this example assume the initial velocity is set at zero for simplicity.
- Step 4: The lifetime of each AN for a particle is calculated. Take the first AN for the first particle as an example. Its lifetime is calculated as follows:

$$l_{11} = 10000/5[(4-1)^2 + (7-10)^2] = 111.11 .$$

The lifetimes of all ANs for all particles are shown in Table 2.

Particle \ AN	1(1, 10)	2(11, 0)	3(8, 7)	4(4, 3)
1(4, 7)	111.11	20.41	100	100
2(9, 5)	22.47	68.97	320	55.17
3(6, 4)	32.79	48.78	123.08	320

Table 2. The lifetimes of all ANs for all particles

- Step 5: The lifetime of the whole sensor system for each particle is calculated as the fitness value. Take the first particle as an example. Its fitness is calculated as follows:

$$Fitness(1) = Min\{l_{11}, l_{12}, l_{13}, l_{14}\} = Min\{111.11, 20.41, 100, 100\} = 20.41.$$

In the same way, the fitness values of all the particles are calculated and shown in Table 3.

Particle	Location	Fitness
1	(4, 7)	20.41
2	(9, 5)	22.47
3	(6, 4)	32.79

Table 3. The fitness values of all the particles

- Step 6: The fitness value of each particle is compared with that of its corresponding *pBest*. If the fitness value of the *i*-th particle is larger than that of $pBest_i$, $pBest_i$ is replaced with the *i*-th particle. In the first generation, the fitness values of all the *pBests* are zero, smaller than those of the particles. The particles are then stored as the new *pBests*. The resulting *pBests* are shown in Table 4, where the field "appearance generation" represents the generation number in which a particle is set as the current *pBest*.

Particle	Location	Fitness	Appearing Generation
1	(4, 7)	20.41	1
2	(9, 5)	22.47	1
3	(6, 4)	32.79	1

Table 4. The *pBests* after the first generation

- Step 7: The best $pBest_i$ among all the particles is chosen as the *gBest*. In this example, $pBest_3$ has the largest fitness value and is set as the *gBest*.
- Step 8: The new velocity of each particle is updated. Assume the inertial weight w is set at 1, the acceleration constant c_1 for particles moving to *pBest* is set at 2, and the acceleration constant c_2 for particles moving to *gBest* is set at 2. Take the first particle as an example to illustrate the step. Its new velocity is calculated as follows:

$$
\begin{aligned}
V_{1x}^{new} &= w \times V_{1x}^{old} + c_1 \times Rand_1() \times (pBest_{1x} - x_{1x}) \\
&\quad + c_2 \times Rand_2() \times (gBest_d - x_{1x}) \\
&= 1 \times 0 + 2 \times 0.5(4-4) + 2 \times 0.25(6-4) \\
&= 1, \text{ and} \\
V_{1y}^{new} &= w \times V_{1y}^{old} + c_1 \times Rand_3() \times (pBest_{1y} - x_{1y}) \\
&\quad + c_2 \times Rand_4() \times (gBest_d - x_{1y}) \\
&= 1 \times 0 + 2 \times 1(7-7) + 2 \times 0.125(4-7) \\
&= -0.75,
\end{aligned}
$$

where the four random numbers generated are 0.5, 0.25, 1 and 0.125, respectively. In the same way, the new velocities of the other two particles can be calculated. The results are shown in Table 5.

Particle	Old Location	Velocity
1	(4, 7)	(1, -0.75)
2	(9, 5)	(-1.2, -0.2)
3	(6, 4)	(0, 0)

Table 5. The new velocities of all the three particles

- Step 9: The position of each particle is updated. Take the first particle as an example. Its new position is calculated as follows:

$$
\begin{aligned}
x_{1x}^{new} &= x_{1x}^{old} + V_{1x}^{new} \\
&= 4 + 1 \\
&= 5, \text{ and} \\
x_{1y}^{new} &= x_{1y}^{old} + V_{1y}^{new} \\
&= 7 + (-0.75) \\
&= 6.25.
\end{aligned}
$$

In the same way, the new positions of all the other two particles can be found. The results are shown in Table 6.

Particle	Old Location	Velocity	New Location
1	(4, 7)	(1, -0.75)	(5, 6.25)
2	(9, 5)	(-1.2, -0.2)	(7.8, 4.8)
3	(6, 4)	(0, 0)	(6, 4)

Table 6. The new positions of all the three particles

- Step 10: Steps 4 to 9 are then repeated until the termination conditions are satisfied. The lifetime evolution along with different generations is shown in Figure 3.

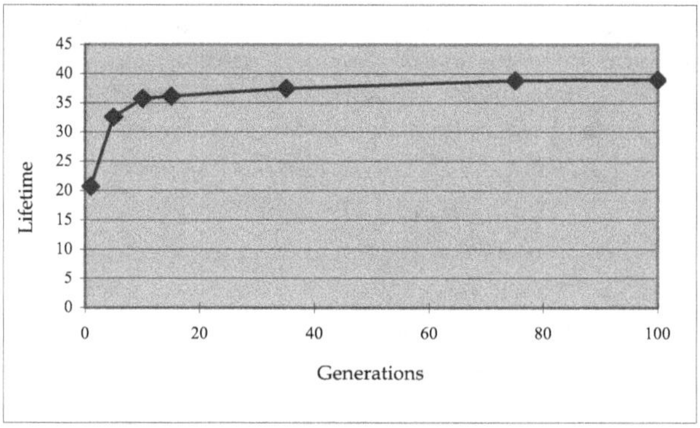

Figure 3. The evolution of the maximal lifetime for the example

6. Experimental Results

Experiments were made to show the performance of the proposed PSO algorithm on finding the optimal positions of base stations. They were performed in C language on an AMD PC with a 2.0GHz processor and 1G main memory and running the Microsoft Window XP operating system. The simulation was done in a two-dimensional real-number space of 1000*1000. That is, the ranges for both x and y axes were within 0 to 1000. The data transmission rate was limited within 1 to 10 and the range of initial energy was limited between 100000000 to 999999999. The data of all ANs, each with its own location, data transmission rate and initial energy, were randomly generated. Note that the data transmission rates and the initial energy amounts of real-life sensors may not fall in the above range. But the proposed approach is still suitable since the lifetime is proportional to the initial energy amount and inversely proportional to the transmission rate.

Experiments were first made to show the convergence of the proposed PSO algorithm when the acceleration constant (c_1) for a particle moving to its *pBest* was set at 2, the acceleration constant (c_2) for a particle moving to its *gBest* was set at 2, the inertial weight (w) was set at 0.6, the distance-independent parameter (a_{j1}) was set at zero, and the distance-dependent parameter (a_{j2}) was set at one. The experimental results of the resulting lifetime along with different generations for 50 ANs and 10 particles in each generation are shown in Figure 4.

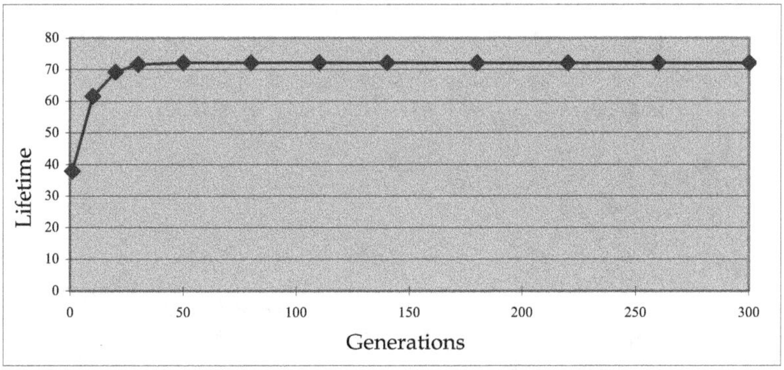

Figure 4. The lifetime for 50 ANs and10 particles

It is easily seen from Figure 4 that the proposed PSO algorithm could converge very fast (below 50 generations). Next, experiments were made to show the effects of different parameters on the lifetime. The influence of the acceleration constant (c_1) for a particle moving to its *pBest* on the proposed algorithm was first considered. The process was terminated at 300 generations. When $w = 1$ and $c_2 = 2$, the nearly optimal lifetimes for 50ANs and 10 particles along with different acceleration constants (c_1) are shown in Figure 5.

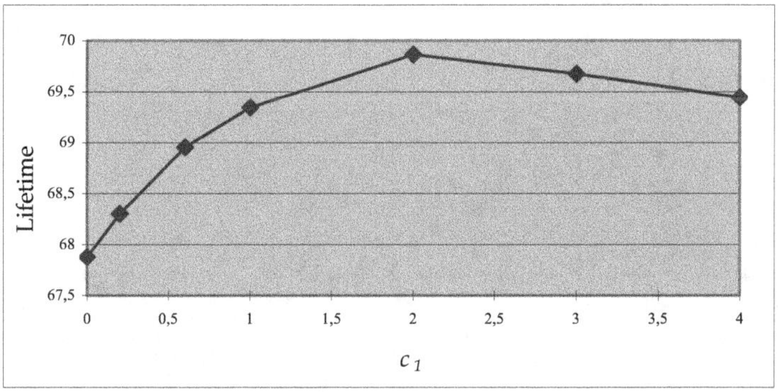

Figure 5. The lifetimes along with different acceleration constants (c_1)

It can be observed from Figure 5 that the lifetime first increased and then decreased along with the increase of the acceleration constant (c_1). When the value of the acceleration constant (c_1) was small, the velocity update of each particle was also small, causing the convergence speed slow. The proposed PSO algorithm might thus not get the optimal solution after the predefined number of generations. On the contrary, when the value of the acceleration constant (c_1) was large, the velocity change would be large as well, causing the particles to move fast. It was then hard to converge. In the experiments, the optimal c_1 value was about 2. Next, experiments were made to show the effects of the acceleration constant (c_2) for a particle moving to its *gBest* on the proposed algorithm. When $w = 1$ and $c_1 = 2$, the experimental results are shown in Figure 6.

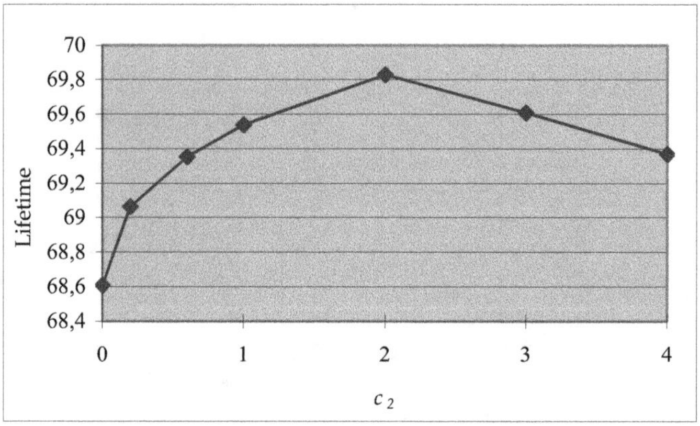

Figure 6. The lifetimes along with different acceleration constants (c_2)

It can be observed from Figure 6 that the lifetime first increased and then decreased along with the increase of the acceleration constant (c_2). The reason was the same as above. In the experiments, the optimal c_2 value was about 2. Next, experiments were made to show the effects of the inertial weight (w) on the proposed algorithm. When c_1 = 2 and c_2 = 2, the experimental results are shown in Figure 7.

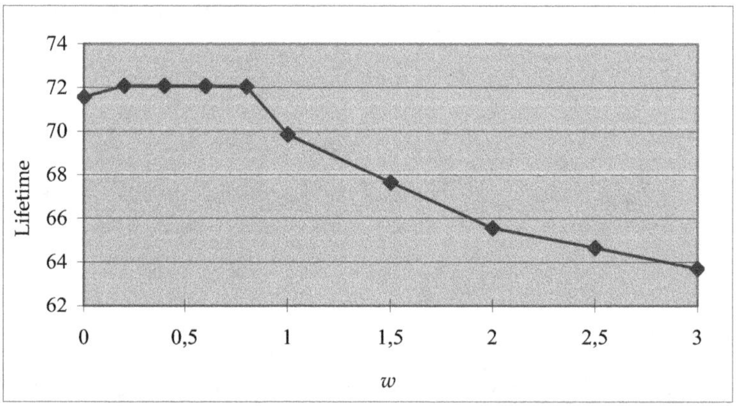

Figure 7. The lifetimes along with different inertial weights (w)

It can be observed from Figure 7 that the proposed algorithm could get good lifetime when the inertial weight (w) was smaller than 0.6. The lifetime decreased along with the increase of the inertial weight (w) when w was bigger than 0.6. This was because when the value of the inertial weight was large, the particles would move fast due to the multiple of the old velocity. It was then hard to converge. Next, experiments were made to show the relation between lifetimes and numbers of ANs. The experimental results are shown in Figure 8.

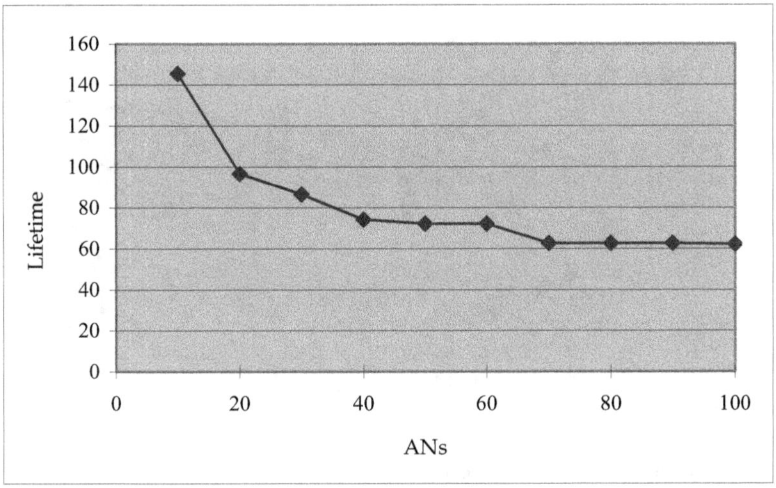

Figure 8. The lifetimes along with different numbers of ANs

It can be seen from Figure 8 that the lifetime decreased along with the increase of the number of ANs. It was reasonable since the probability for at least an AN in the system to fail would increase when the number of ANS grew up. The execution time along with different numbers of ANs is shown in Figure 9.

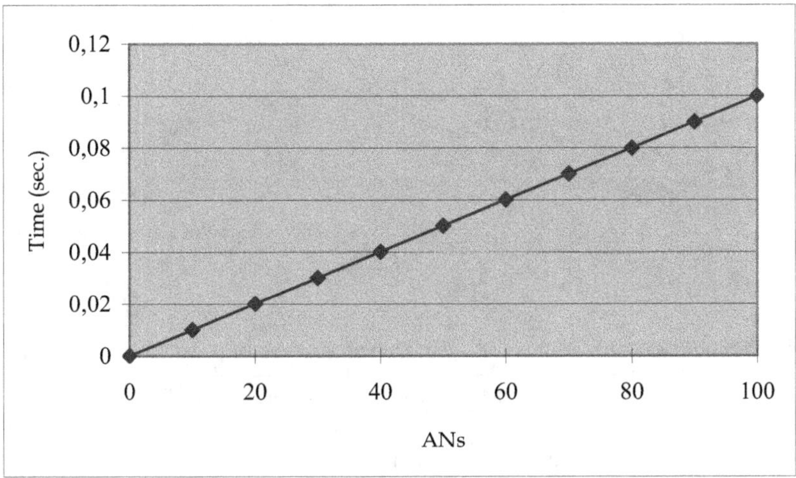

Figure 9. The execution time along with different numbers of ANs

It can be observed from Figure 9 that the execution time increased along with the increase of numbers of ANs. The relation was nearly linear. Experiments were then made to show the relation between lifetimes and numbers of particles for 50 ANs and 300 generations. The internal weight was set at 1. The experimental results are shown in Figure 10.

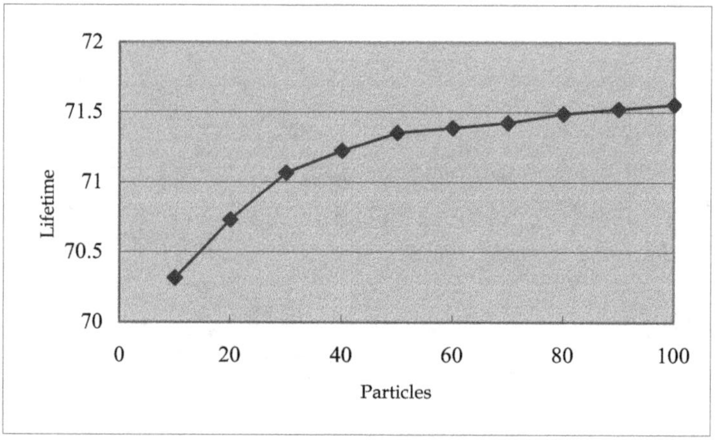

Figure 10. The lifetimes along with different numbers of particles

It can be seen from Figure 10 that the lifetime increased along with the increase of numbers of particles for the same number of generations. The execution time along with different numbers of particles for 300 generations is shown in Figure 11.

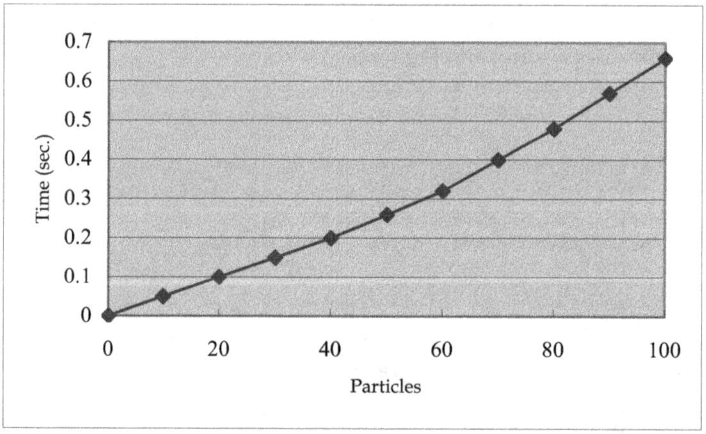

Figure 11. The execution time along with different numbers of particles for 50 ANs

Method	Lifetime
The proposed PSO algorithm	72.0763
The exhaustive grid search (grid size = 1)	72.0048
The exhaustive grid search (grid size = 0.1)	72.0666
The exhaustive grid search (grid size = 0.01)	72.0752

Table 7. A lifetime comparison of the PSO approach and the exhaustive grid searc

It can be observed from Figure 11 that the execution time increased along with the increase of numbers of particles. The relation was nearly linear. This was reasonable since the execution time would be approximately proportional to the number of particles.

Note that no optimal solutions can be found in a finite amount of time since the problem is NP-hard. For a comparison, an exhaustive search using grids was used to find nearly optimal solutions. The approach found the lifetime of the system when a BS was allocated at any cross-point of the grids. The cross-point with the maximum lifetime was then output as the solution. A lifetime comparison of the PSO approach and the exhaustive search with different grid sizes are shown in Table 7.

It can be observed from Table 7 that the lifetime obtained by our proposed PSO algorithm was not worse than those by the exhaustive grid search within a certain precision. The lifetime by the proposed PSO algorithm was 72.0763, and was 72.0048, 72.0666 and 72.0752 for the exhaustive search when the grid size was set at 1, 0.1 and 0.01, respectively. For the exhaustive grid search, the smaller the grid size, the better the results.

7. Conclusion

In wireless sensor networks, minimizing power consumption to prolong network lifetime is very crucial. In this paper, a two-tiered wireless sensor networks has been considered and an algorithm based on particle swarm optimization (PSO) has been proposed for general power-consumption constraints. The proposed approach can search for nearly optimal BS locations in heterogeneous sensor networks, where ANs may own different data transmission rates, initial energies and parameter values. Finding BS locations is by nature very similar to finding the food locations originated from PSO. It is thus very easy to model such a problem by the proposed algorithm based on PSO. Experiments have also been made to show the performance of the proposed PSO approach and the effects of the parameters on the results. From the experimental results, it can be easily concluded that the proposed PSO algorithm converges very fast when compared to the exhaustive search. It can also be easily extended to finding multiple base stations.

8. Acknowledgement

This research was supported by the National Science Council of the Republic of China under contract NSC 97-2221-E-390-023. We also like to thank Mr. Cheng-Hsi Wu for his help in making part of the experiments.

9. References

J. Chou, D. Petrovis and K. Ramchandran, A distributed and adaptive signal processing approach to reducing energy consumption in sensor networks, *The 22nd IEEE Conference on Computer Communications (INFOCOM)*, pp. 1054-1062, San Francisco, USA, March, 2003. [1]

R. C. Eberhart and J. Kennedy, A new optimizer using particles swarm theory, *The Sixth International Symposium on Micro Machine and Human Science*, pp. 39-43, Nagoya, Japan, October, 1995. [2]

R. C. Eberhart and J. Kennedy, Particle swarm optimization, *The IEEE International Conference on Neural Networks*, Vol. 4, pp. 1942-1948, Perth, Australia, November, 1995. [3]

Z. L. Gaing, Discrete particle swarm optimization algorithm for unit commitment, *The IEEE Power Engineering Society General Meeting*, Toronto, Canada, July, 2003. [4]

W. Heinzelman, J. Kulik, and H. Balakrishnan, Adaptive protocols for information dissemination in wireless sensor networks, *The Fifth ACM International Conference on Mobile Computing and Networking*, pp. 174-185, Seattle, USA, August, 1999. [5]

W. Heinzelman, A. Chandrakasan, and H. Balakrishnan, Energy-efficient communication protocols for wireless microsensor networks, *The Hawaiian International Conference on Systems Science,* Hawaii, USA, January, 2000. [6]

C. Intanagonwiwat, R. Govindan and D. Estrin, Directed diffusion: a scalable and robust communication paradigm for sensor networks, *The ACM International Conference on Mobile Computing and Networking*, Boston, USA, August, 2000. [7]

Vikas Kawadia and P. R. Kumar, Power control and clustering in ad hoc networks, *The 22nd IEEE Conference on Computer Communications (INFOCOM),* San Francisco, USA, March, 2003. [8]

N. Li, J. C. Hou and L. Sha, Design and analysis of an mst-based topology control algorithm, *The 22nd IEEE Conference on Computer Communications (INFOCOM)*, San Francisco, USA, March, 2003. [9]

S. Lee, W. Su and M. Gerla, Wireless ad hoc multicast routing with mobility prediction, *Mobile Networks and Applications*, Vol. 6, No. 4, pp. 351-360, 2001. [10]

D. Niculescu and B. Nath, Ad hoc positioning system (APS) using AoA, *The 22nd IEEE Conference on Computer Communications (INFOCOM)*, pp. 1734-1743, San Francisco, USA, March, 2003. [11]

J. Pan, Y. Hou, L. Cai, Y. Shi and X. Shen, Topology control for wireless sensor networks, *The Ninth ACM International Conference on Mobile Computing and Networking*, pp. 286-299, San Diego, USA, September, 2003. [12]

J. Pan, L. Cai, Y. T. Hou, Y. Shi and S. X. Shen, Optimal base-station locations in two-tiered wireless sensor networks, *IEEE Transactions on Mobile Computing*, Vol. 4, No. 5, pp. 458-473, 2005. [13]

Y. Q. Qin, D. B. Sun, N. Li and Y. G. Cen, Path planning for mobile robot using the particle swarm optimization with mutation operator, *The IEEE Third International Conference on Machine Learning and Cybernetics*, pp. 2473-2478, Shanghai, China, August, 2004. [14]

V. Rodoplu and T. H. Meng, Minimum energy mobile wireless networks, *IEEE Journal on Selected Areas in Communications,* Vol. 17, No. 8, 1999. [15]

R. Ramanathan and R. Hain, Topology control of multihop wireless networks using transmit power adjustment, *The 19th IEEE Conference on Computer Communications (INFOCOM)*, Tel Aviv, Israel, March, 2000. [16]

Y. Shi and R. C. Eberhart, A modified particle swarm optimizer, *The IEEE International Conference on Evolutionary Computation*, pp. 69-73, Anchorage, USA, May, 1998, [17]

A. Stacey, M. Jancic and I. Grundy, Particle swarm optimization with mutation, *The IEEE Congress on Evolutionary Computation*, pp. 1425-1430, Canberra, Australia, December, 2003. [18]

D. Tian and N. Georganas, Energy efficient routing with guaranteed delivery in wireless sensor networks, *The IEEE Wireless Communication and Networking Conference,* Orleans, USA, March, 2003. [19]

F. Ye, H. Luo, J. Cheng, S. Lu and L. Zhang, A two-tier data dissemination model for large scale wireless sensor networks, *The ACM International Conference on Mobile Computing and Networking,* Atlanta, USA, September, 2002. [20]

4

Particle Swarms in Statistical Physics*

Andrei Băutu [1] and Elena Băutu [2]
[1] *"Mircea cel Batrân" Naval Academy,* [2] *"Ovidius" University*
Romania

1. Introduction

Statistical physics is the area of physics which studies the properties of systems composed of many microscopic particles (like atoms and molecules). When combined, the interactions between these particles produce the macroscopic features of the systems. The systems are usually characterized by a very large number of variables and the limited possibilities for observing the properties of the components of the system. For these reasons, solving problems arisen in Statistical Physics with analytical approaches is usually ineffective and sometimes impossible. However, statistical approaches (such as Monte Carlo simulation) can provide acceptable approximations for solutions of these problems. Moreover, recent studies showed that nature inspired metaheuristics (like Genetic Algorithms, Evolutionary Strategies, Particle Swarm Optimization, etc) can also be used to simulate, analyse, and optimize such systems, providing fast and accurate results. Apart from physical implications, problems from Statistical Physics are also important in fields like biology, chemistry, mathematics, communications, economy, sociology, etc.

We will present two important problems from Statistical Physics and discuss how one can use Particle Swarm Optimization (PSO) to tackle them. First, we will discuss how the real-valued version of PSO can be used to minimize the energy of a system composed of repulsive point charges confined on a sphere. This is known as the Thomson problem and it is included in Stephen Smale's famous list of 18 unsolved mathematical problems to be solved in the 21st century. This problem also arises in biology, chemistry, communications, economy, etc.

Latter on, we will discuss how the binary version of PSO can be used to search ground states of Ising spin glasses. Spin glasses are materials that simultaneously present ferromagnetic and anti-ferromagnetic interactions among their atoms. A ground state of a spin glass is a configuration of the system in which this has the lowest energy possible. Besides its importance for Statistical Physics, this problem has applications in neural network theory, computer science, biology, etc.

2. The Basics of Particle Swarm Optimization

Particle Swarm Optimization (PSO) is a metaheuristic inspired by the behaviour of social creatures, which interact between them in order to achieve a common goal. Such behaviour

---

* This work was in part supported by CNCSIS grants TD 254/2008 and TD 199/2007.

can be noticed in flocks of birds searching for food or schools of fish trying to avoid predators (Eberhart & Kennedy, 1995).

The philosophy of PSO is based on the evolutionary cultural model, which states that in social environments individuals have two learning sources: individual learning and cultural transmission (Boyd & Richerson, 1988). Individual learning is an important feature in static and homogeneous environments, because one individual can learn many things about the environment from a single interaction with it. However, if the environment is dynamic or heterogeneous, then that individual needs many interactions with the environment before it gets to know it. Because a single individual might not get enough chances to interact with such environment, cultural transmission (meaning learning from the experiences of others) becomes a requisite, too. In fact, individuals that have more chances to succeed in achieving their goals are the ones that combine both learning sources, thus increasing their gain in knowledge.

In order to solve any problem with PSO, we need to define a fitness function which will be used to measure the quality of possible solutions for that problem. Then, solving the original problem is equivalent to optimizing parameters of the fitness function, such that we find one of its minimum or maximum values (depending on the fitness function). By using a set of possible solutions, PSO will optimize our fitness function, thus solving our original problem. In PSO terms, each solution is called a particle and the set of particles is called a swarm. Particles gather and share information about the problem in order to increase their quality and hopefully become the optimum solution of the problem. Therefore, the driving force of PSO is the collective swarm intelligence (Clerc, 2006).

The fitness function generates a problem landscape in which each possible solution has a corresponding fitness value. We can imagine that similar to birds foraging in their environment, the PSO particles move in this landscape searching locations with higher rewards and exchanging information about these locations with their neighbours. Their common goal is to improve the quality of the swarm. During the search process the particles change their properties (location, speed, memory, etc) to adapt to their environment.

3. Particle Swarm Optimization and the Thompson problem

In 1904, while working on his plum pudding model of the atom, the British physicist Joseph John Thomson stated the following problem: what is the minimum energy configuration of N electrons confined on the surface of a sphere? Obviously, each two electrons repel each other with a force given by Coulomb's law:

$$F = \frac{q_e^2}{4\pi\varepsilon_0}\frac{1}{d^2}, \tag{1}$$

where ε_0 is the electric constant of vacuum, q_e is the charge on a single electron, and d is the distance between the two electrons. Because of these forces, each electron will try to get as far as possible from the others. However, being confined on the surface of the sphere, they will settle for a system configuration with minimum potential energy. The potential energy of a system with N electrons is:

$$U(S)=\frac{q_e^2}{4\pi\varepsilon_0}\sum_{j=1}^{N-1}\sum_{i=j+1}^{N}\frac{1}{d_{ij}}, \tag{2}$$

where we consider the electrons numbered in some random fashion, and d_{ij} is the distance between electrons i^{th} and j^{th} in the system configuration S. Any configuration with minimum potential energy is called a "ground state" of the system.

Over the years, various generalizations for the Thomson problem have also been studied from different aspects. The most common generalization involves interactions between particles with arbitrary potentials. Bowick studied a system of particles constrained to a sphere and interacting by a $d^{-\gamma}$ potential, with $0<\gamma<2$ (Bowick et al, 2002). Travesset studied the interactions of the particles on topologies other than the 2-sphere (Travesset, 2005). Levin studied the interactions in a system with $N-1$ particles confined on the sphere and 1 particle fixed in the centre of the sphere (Levin & Arenzon, 2003). In general, finding the ground state of a system of N repulsive point charges constrained to the surface of the 2-sphere is a long standing problem, which was ranked 7 in Stephen Smale's famous list (Smale, 2000) of 18 unsolved mathematical problems to be solved in the 21st century, along with other famous problems, like the Navier-Stokes equations, Hilbert's sixteenth problem and the P=NP problem.

Apart from physics, the Thomson problem arises in various forms in many other fields: biology (determining the arrangements of the protein subunits which comprise the shells of spherical viruses), telecommunications (designing satellite constellations, selecting locations for radio relays or access points for wireless communications), structural chemistry (finding regular arrangements for proteins S-layers), mathematics, economy, sociology, etc. From an optimization point of view, the Thomson problem is of great interest to computer scientists also, because it provides an excellent test bed for new optimization algorithms, due to the exponential growth of the number of minimum energy configurations and to their characteristics.

The Thomson problem can be solved exactly for small values of N point charges on the surface of a sphere or a torus. However, for large values of $N>8$, exact solutions are not known. The configurations found so far for such values display a great variety of geometrical structures. The best known solutions so far for such systems were identified with numerical simulations, using methods based on Monte Carlo simulations, evolutionary algorithms, simulated annealing, etc (Carlson et al., 2003; Morris et al., 1996; Perez-Garrido et al., 1996; Pang, 1997). PSO for the Thompson problem was first introduced in (Băutu & Băutu, 2007).

We will present in the following how the real-valued version of PSO can be used to tackle the Thomson problem. In order to avoid confusion, we will use the term "point charges" to refer to physical particles on the sphere (electrons, for example) and "particles" to refer to the data structures used by the PSO algorithm.

As mentioned in the previous section, in order to use a PSO algorithm we need to define a function that will measure the quality of solutions encoded by particles. One can think of many such functions for the Thompson problem, but a simple and quick solution is to use the potential energy of a system. We can save some computation time if we ignore the physical constants and use a simplified version of (2) for our fitness function:

$$F(P) = \sum_{j=1}^{N-1} \sum_{i=j+1}^{N} \frac{1}{d_{ij}} \tag{3}$$

where N is the number of point charges in the system, d_{ij} is the Euclidian distance between point charges i and j, encoded by the particle P. If we represent our system configuration in 3D space using a Cartesian coordinate system, then we need to handle $3N$ real values, for the values on the Ox, Oy and Oz axis of each particle (see Figure 1). We will also need to explicitly enforce the sphere surface constraints which require additional computation time.

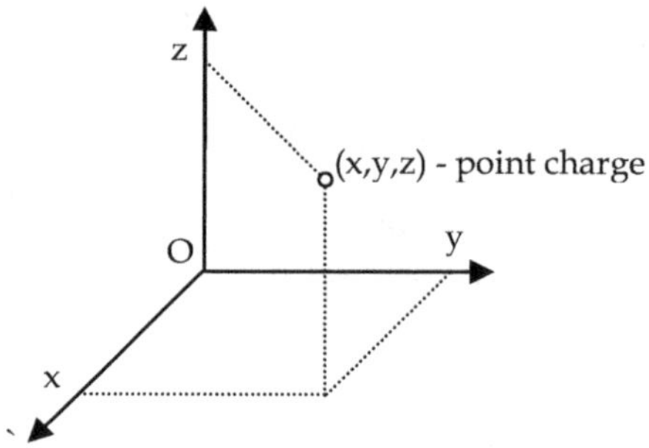

Figure 1. Point charge represented in 3D Cartesian coordinate system

The memory requirements can be reduced and the computation overhead for constraint enforcing can be avoided, if we scale our system to the unit sphere and represent its configuration using a Spherical coordinate system. In this way, the sphere surface constraint is implicitly enforced and since r is constant, the system configuration is encoded with only $2N$ real values, representing the azimuth ϕ and elevation θ angles (see Figure 2).

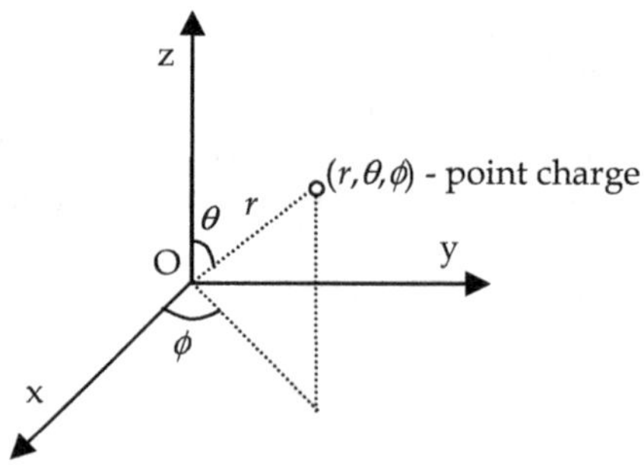

Figure 2. Point charge represented in 3D Spherical coordinate system

In this case, the distance between point charges i and j located on the surface of the unit sphere is

$$d_{ij} = \sqrt{2 - 2[\cos\phi_i \cos\phi_j + \sin\phi_i \sin\phi_j \cos(\theta_i - \theta_j)]}\ , \tag{4}$$

where $\phi_1, \phi_2 \in [-\pi, \pi]$ is the azimuth angle and $\theta_i, \theta_j \in [-\pi/2, \pi/2]$ is the elevation angle. Thus, PSO particles move in the search space $[0,1]^{2N}$ and the location $x \in [0,1]^{2N}$ of a particle decodes into a system configuration with:

$$\phi_i = 2\pi x_{2i-1} - \pi \tag{5}$$

$$\theta_i = \pi x_{2i} - \pi/2 \tag{6}$$

With this setup in place, the PSO algorithm begins with a swarm of particles randomly scattered around the search space. This generic initialization method could be replaced with a problem specific one (spherical initialization, for example). Each particle has a set of neighbours with which it will exchange information. An iterative process begins, which updates the properties of the particles. On each iteration each particle use the information from its own memory and the information gathered from its neighbours to update its properties. The equation used for updating the speed is:

$$v_t = \omega v_{t-1} + \phi_1 R_1 (p_{t-1} - x_{t-1}) + \phi_2 R_2 (g_{t-1} - x_{t-1}), \tag{7}$$

where v_t is the speed at iteration t, x_t is the location of the particle at iteration t, p_t is the best location the particle has found until iteration t, g_t is the best location the neighbours of the particle found up to the iteration t. The individual learning and cultural transmission factors (ϕ_1 and ϕ_2) control the importance of the personal and neighbour's experience on the search process. Note that although they share the same notation, these are parameters of the algorithm and are distinct and not related to the azimuth angles of the point charges. Because the importance of individual learning and cultural transmission is unknown, the learning factors are weighted by random values $R_1, R_2 \in [0,1)$. Usually the speed is bounded by some $\mathrm{v}_{\max}$ parameter to prevent it from increasing too much because of these random values.

With the updated speed vector and the old position of the particle, the new position is computed with:

$$\mathrm{x}_{\mathrm{t}} = x_{t-1} + v_t \Delta t, \tag{8}$$

for $\Delta t = 1$ iteration.

Based on the previous discussion, the PSO algorithm used for the Thomson problem is summarized in Figure 3. The algorithm is very simple and requires basic programming skills to be implemented in any programming language. It has many parameters that can be tuned in order to achieve high performance results. The tuning process of these parameters is beyond the purpose of this chapter. For now, let's consider the following setup: $\omega = 0.9$ – will allow the algorithm to avoid rapid changes in the trajectories of the particles; $\phi_1 = \phi_2 = 2$ – gives equal weight to individual and social learning; iterations = 500 – for small and medium size systems, this should be enough for the particles to discover and focus on good solutions; $M = 2N$ – increases the swarm size with the size of the system.

```
Initialize M random particles
for t = 1 to iterations
    for each particle
        Update v_t according to (7)
        Update x_t according to (8)
        Decode x_t using (5) and (6)
        Evaluate x_t using (3) and (4)
        Update p_t and g_t according to their definition
    next
next
return solution from the particle with smaller fitness
```

Figure 3. PSO algorithm for the Thomson problem

Performing 10 runs of the algorithm from Figure 3 for systems with different sizes, we obtained the results presented in Table 1:

N	Minimum known energy	Energy of PSO solution
2	0.500000000	0.500000000
3	1.732050808	1.732050808
4	3.674234614	3.674234614
5	6.474691495	6.474691495
6	9.985281374	9.985281374
7	14.452997414	14.452987365
8	19.675287861	19.675287861
9	25.759986531	25.759986599
10	32.716949460	32.717003657
15	80.670244114	80.685310397
20	150.881568334	150.953443814
25	243.812760299	243.898092955
30	359.603945904	359.901863399
35	498.569872491	499.018395878
40	660.675278835	661.666117852
45	846.188401061	847.129739052
50	1055.182314726	1056.517970873

Table 1. Minimum energies for Thomson problem found in experiments

From the results in Table 1, one can see that this simple PSO algorithm can provide high quality estimates for the ground states of various instances of the Thomson problem. The algorithm can be further improved not only in its parameters, but also in its structure (using a more advanced initialization method, for example). Obviously, the Particle Swarm Optimization algorithm can be applied for generalized forms of Thomson problem and other related problems, not only from Statistical Physics, but other domains, too.

4. Binary Particle Swarm Optimization and Ising Spin Glasses

Matter is composed of atoms and each atom carries a spin, meaning the magnetic moment of the microscopic magnetic field around the atom generated by the motion of the electrons around its nucleus.

If we heat a metal object higher than the Curie point of its material, the object will loose its ferromagnetic properties and become paramagnetic. At that point, the spins of the atoms change randomly so erratic that at any time they can point with equal probability to any possible direction. In this case, the individual microscopic magnetic fields generated by the spins cancel each other out, such that there is no macroscopic magnetic field (Huang, 1987).

When the temperature is lower than the Curie point, in some metals (iron and nickel, for example) the spins of the atoms tend to be polarized in the same direction, producing a measurable macroscopic magnetic field. This is called "ferromagnetic" behaviour. By contrast, below the Curie point, in spin glasses only some pairs of neighbouring spins prefer to be aligned, while the others prefer to be anti-aligned, resulting two types of interactions between atoms: ferromagnetic and anti-ferromagnetic. Because of this mix of interactions, these systems are called disordered (den Hollander & Toninelli, 2005).

In the past, condensed matter physics has focused mainly on ordered systems, where symmetry and regularity lead to great mathematical simplification and clear physical insight. Over the last decades, spin glasses became a thriving area of research in condensed matter physics, in order to understand disordered systems. Spin glasses are the most complex kind of condensed state encountered so far in solid state physics (De Simone et al., 1995). Some examples of spin glasses are metals containing random magnetic impurities (called disordered magnetic alloys), such as gold with small fractions of iron added (AuFe).

Apart from their central role in Statistical Physics, where they are the subject of extensive theoretical, experimental and computational investigation, spin glasses also represent a challenging class of problems for testing optimization algorithms. The problem is interesting because of the properties of spin glass systems, such as symmetry or large number of plateaus (Pelikan & Goldberg, 2003).

From an optimization point of view, the main objective is to find the minimum energy for a given spin glass system (Hartmann, 2001; Pelikan & Goldberg, 2003; Fischer, 2004; Hartmann & Weigt, 2005). System configurations with the lowest energy are called ground states and thus the problem of minimizing the energy of spin glass instances can be formulated as the problem of finding ground states of these instances (Pelikan et al., 2006). The main difficulties when searching for ground states of spin glasses come from the many local optima in the energy landscape which are surrounded by high-energy neighbouring configurations (Pelikan & Goldberg, 2003).

The Ising model is a simplified description of ferromagnetism, yet it is extremely important because other systems can be mapped exactly or at least approximately to it. Its applications range from neural nets and protein folding to flocking birds, beating heart cells and more. It was named after the German physicist Ernst Ising who first discussed it in 1925, although it was suggested in 1920 by his Ph.D. advisor, Wilhelm Lenz. Ising used it as a mathematical model for phase transitions with the goal of explaining how long-range correlations are generated by local interactions.

The Ising model can be formulated for any dimension in graph-theoretic terms. Let us consider a spin glass system with N spins and no external magnetic field. The interaction graph $G=(V,E)$ associated with the system has the vertex set $V=\{v_1,\ldots,v_N\}$. Each vertex

$i \in V$ can be in one of two states $S_i \in \{-1,1\}$. Edges in this graph represent bonds between adjacent atoms in the spin glass system. Each edge $ij \in E$ has assigned a coupling constant, denoted by $J_{ij} \in \{-J, J\}$; an edge exists between vertices i and j if the interaction between atoms i and j is not zero. In the classic model, this graph is a standard "square" lattice in one, two, or three dimensions. Therefore, each atom has two, four, or six nearest neighbours, respectively (see Figure 4). However, various papers present research done on larger dimensions (Hartmann, 2001).

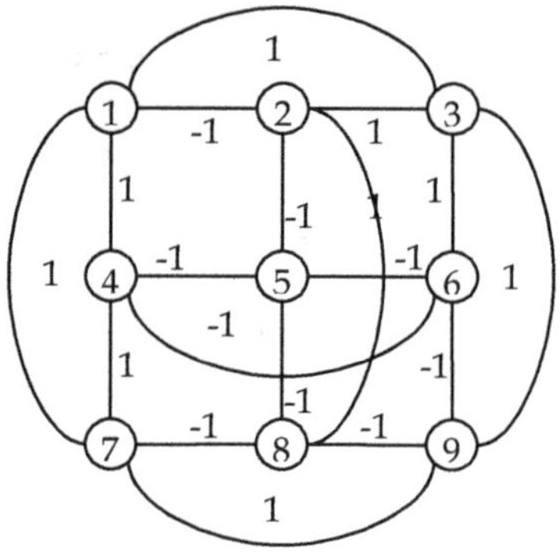

Figure 4. Two dimensional Ising spin glass system

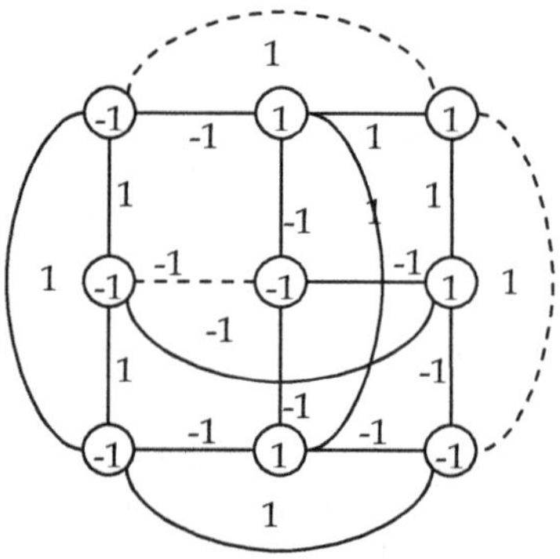

Figure 5. Ground state of the system from Figure 4 (values inside circles represent the states of the spins; dashed lines represent unsatisfied bonds)

For a system configuration S, the interaction between neighbouring vertices i and j contributes an amount of $-J_{ij}S_iS_j$ to the total energy of the system, expressed as the Hamiltonian:

$$H(S) = -\sum_{ij \in E} J_{ij} S_i S_j \,. \tag{9}$$

The sign of J_{ij} gives the nature of the interaction between neighbours i and j. If J_{ij} is positive, the interaction is ferromagnetic. Having the two neighbours in the same state ($S_i = S_j$) decreases the total energy. If J_{ij} is negative, the interaction between neighbours i

and j is anti-ferromagnetic. The decrease in total energy is obtained if they have opposite states. When all coupling constants are positive (or negative), a lowest-energy configuration is obtained when all vertices have the same state. This is the case of ferromagnetic materials. When the coupling constants are a mix of positive and negative values, as is the case for spin glasses, finding the "ground state" is a very difficult problem. A ground state of the system from Figure 4 is presented in Figure 5.

The two-dimensional Ising model of ferromagnetism has been solved exactly by Onsager (Onsager, 1944). The most common configurations in the literature are 2D Ising spin glasses on a grid with nearest neighbour interactions. In the case of no periodic boundary conditions and no exterior magnetic field, the problem reduces to finding a minimum weight cut in a planar graph for which polynomial time algorithms exist (Orlova & Dorfman, 1972; Goodman & Hedetniemi, 1973; Hadlock, 1975). Barahona showed that finding a ground state for the three-value coupling constant ($J_{ij} \in \{-1,0,1\}$) on a cubic grid is equivalent to finding a maximum set of independent edges in a graph for which each vertex has degree 3 (Barahona, 1982). He also showed that computing the minimum value of the Hamiltonian of a spin glass with an external magnetic field,

$$H(S) = -\sum_{ij \in E} J_{ij} S_i S_j - h \sum_{i \in V} S_0 S_i \, , \tag{10}$$

is equivalent to solving the problem of finding the largest set of disconnected vertices in a planar, degree-3 graph. This means that finding ground states for three-dimensional spin glasses on the standard square lattice and for planar spin glasses with an external field are NP-complete problems. Istrail showed that the essential ingredient for the NP-completeness of the Ising model is the non planarity of the graph (Istrail, 2000).

Particle Swarm Optimization was introduced as a technique for numerical optimization and has proved to be very efficient on many real-valued optimization problems. Because finding the ground state of a spin glass system in the Ising model is a combinatorial problem, we need to apply a modified version of PSO. We will use the binary version of PSO (Kennedy & Eberhard, 1997). In this case, the ith component of the position vector of a particle encodes the state of the ith spin in the system (0 means down, 1 means up), while the ith component of the velocity vector determines the confidence of the particle that the ith spin should be up.

On each iteration of the search process, each particle updates its velocity vector (meaning its confidence that the spins should be up) using (7). After that, the particle's position vector (meaning its decision about spins being up or down) it updated using the component-wise formula:

$$x_{ti} = \begin{cases} 1, \text{if } R < \left(1 + \exp(-v_{ti})\right)^{-1} \\ 0, \text{otherwise} \end{cases} \tag{11}$$

where $R \in [0,1)$ is a random value. Once a particle's position is known, its profit can be computed by:

$$F(x_t) = -\sum_{1 \le i < j \le N} J_{ij} (2x_{ti} - 1)(2x_{tj} - 1), \tag{12}$$

With such fitness function, lower values indicate better solutions. Obviously, this fitness function is inspired by the Hamiltonian given in (9) and can be adapted easily to external magnetic field environments using (10).

Based on the previous discussion, the Binary PSO algorithm used for the Ising spin glass problem is presented in Figure 6. A more advanced PSO algorithm for this problem is described in (Băutu et al., 2007). It combines the PSO algorithm with a local optimization technique which allows the resulting hybrid algorithm to fine tune candidate solutions.

```
Initialize M random particles
for t = 1 to iterations
    for each particle
        Update v_t according to (7)
        Update x_t according to (11)
        Evaluate x_t using (12)
        Update p_t and g_t according to their definition
    next
next
return solution from the particle with smaller fitness
```

Figure 6. PSO algorithm for the Ising spin glass problem

In order to test this algorithm, one can use a spin grass system generator, like the Spin Glass Server (SGS). SGS can be used to solve exactly 2D and 3D systems with small sizes or to generate systems for testing. It is available online at http://www.informatik.uni-koeln.de/ls_juenger/research/sgs/sgs.html.

N	SGS minimum energy per spin	PSO minimum energy per spin
64 / 3D	-1.6875	-1.6875
64 / 3D	-1.7500	-1.7500
64 / 3D	-1.8750	-1.8750
125 / 3D	-1.7040	-1.6720
125 / 3D	-1.7680	-1.7360
125 / 3D	-1.7360	-1.7040

Table 2. Minimum energies for Ising spin glasses found in experiments

Table 2 presents the energy per spin values obtained for 3D systems of 4x4x4 and 5x5x5 spins using (13). They will give you an idea about the performance of the binary PSO on this type of problems. The actual values depend on the spin system for which the algorithm is used.

$$E(x_t) = \frac{F(x_t)}{N} \tag{13}$$

The results from table 2 were obtained without any tuning of the PSO parameters: the individual and social learning factors are $\phi_1 = \phi_2 = 2$ and the inertia factor is $\omega = 0.9$. The number of iterations is twice the number of spins, and the number of particles is three times the number of spins. SGS provides the minimum energy for these systems using a branch-and-cut algorithm (De Simone et al., 1995).

5. Conclusions

This chapter presented the basic traits of Particle Swarm Optimization and its applications for some well known problems in Statistical Physics. Recent research results presented in the literature for these problems prove that PSO can find high quality solutions in reasonable times (Băutu et al, 2007; Băutu & Băutu, 2008). However, many questions are still open: how do the parameters setups relate to the problems tackled? how can we improve the basic PSO to get state of the are results? how can we tackle very large size systems?

6. References

Barahona, F. (1982). On the Computational Complexity of Ising Spin Glass Models, *Journal of Physics A: Mathematical and General*, 15(10), Oct. 1982, pp. 3241-3253

Băutu, A., Băutu, E. & Luchian, H. (2007). Particle Swarm Optimization Hybrids for Searching Ground States of Ising Spin Glasses, *Proceedings of the Ninth International Symposium on Symbolic and Numeric Algorithms for Scientific Computing*, Timisoara, Romania, pp. 415-418, ISBN 0-7695-3078-8, IEEE Computer Society

Băutu, A. & Băutu, E. (2007). Energy Minimization Of Point Charges On A Sphere With Particle Swarms, *7th International Balkan Workshop on Applied Physics*, Constantza, Jul. 2007

Băutu, A. & Băutu, E. (2008). Searching Ground States of Ising Spin Glasses with Genetic Algorithms and Binary Particle Swarm Optimization, *Nature Inspired Cooperative Strategies for Optimization (NICSO 2007)*, vol. 129, May 2008, pp. 85-94, ISBN 978-3-540-78986-4, Springer, Berlin

Bowick, M., Cacciuto, A., Nelson, D.R. & Travesset, A. (2002). Crystalline Order on a Sphere and the Generalized Thomson Problem, *Physical Review Letters*, 89(18), Oct. 2002, pp. 185502, ISSN 0031-9007

Boyd, R. & Richerson, P.J. (1988). *Culture and the Evolutionary Process*, University of Chicago Press, ISBN 0226069338, Chicago

Carlson, J., Chang, S.Y., Pandharipande, V.R. & Schmidt, K.E. (2003). Superfluid Fermi Gases with Large Scattering Length, *Physical Review Letters*, 91(5), Aug. 2003, pp. 050401, ISSN 0031-9007

Clerc, M. (2006). *Particle Swarm Optimization*, Hermes Science Publishing Ltd., ISBN 1905209045, London

De Simone, C., Diehl, M., Junger, M., Mutzel, P. Reinelt, G., & Rinaldi, G. (1995). Exact ground states of Ising spin glasses: New experimental results with a branch and cut algorithm, *Journal of Statistical Physics*, 80(2), Jul. 1995, pp. 487-496, ISSN 0022-4715

Eberhart, R. & Kennedy, J. (1995). A new optimizer using particle swarm theory, *Proceedings of the Sixth International Symposium on Micro Machine and Human Science*, pp. 39-43, ISBN 0-7803-2676-8, Nagoya, Oct. 1995, IEEE Press, Piscataway

Fischer, S. (2004). A Polynomial Upper Bound for a Mutation-Based Algorithm on the Two-Dimensional Ising Model, *Proceedings of GECCO-2004, Part I*, pp. 1100-1112, ISBN 3-540-22344-4, Genetic and Evolutionary Computation, Jun. 2004, Springer, Berlin

Goodman, S. & Hedetniemi, S. (1973). Eulerian walks in graphs, *SIAM Journal on Computing*, 2(2), Mar. 1973, pp. 16-27, SIAM, ISSN 0097-5397

Hadlock, F. (1975) Finding a Maximum Cut of a Planar Graph in Polynomial Time, *SIAM Journal on Computing*, 4(3), Sep.1975, pp. 221-225, SIAM, ISSN 0097-5397

Hartmann, A.K. (2001). Ground-State Clusters of Two-, Three- and Four-dimensional ±J Ising Spin Glasses, *Physical Review E*, 63(2), Jan. 2001, pp. 016106, ISSN 0031-9007

Hartmann, A.K. & Weigt, M. (2005). *Phase Transitions in Combinatorial Optimization Problems: Basics, Algorithms and Statistical Mechanics*, Wiley, ISBN 3-527-40473-5, New York

den Hollander, F., Toninelli, F. (2005). Spin glasses: A mystery about to be solved, *Eur. Math. Soc. Newsl.*, vol. 56, pp. 13-17, 2005

Huang, K. (1987) *Statistical Mechanics, 2nd ed.*, Wiley, ISBN 0471815187, New York

Istrail, S. (2000). Universality of Intractability of the Partition Functions of the Ising Model Across Non-Planar Lattices, *Proceedings of the STOC00*, pp. 87-96, ISBN 1-58113-184-4, 32nd ACM Symposium on the Theory of Computing, May 2000, ACM Press

Kennedy, J. & Eberhart, R.C. (1997). A discrete binary version of the particle swarm algorithm, *Proceedings of the World Multiconference on Systemics, Cybernetics and Informatics*, vol. 5, IEEE International Conference on Systems, Man, and Cybernetics, pp. 4104-4109, ISBN 0-7803-4053-1, IEEE Press, Piscataway

Levin, Y. & Arenzon, J.J. (2003) Why charges go to the surface: A generalized Thomson problem, *Europhysics Letters*, 63(3), Aug. 2003, pp. 415-418, ISSN 0295-5075

Morris, J.R., Deaven, D.M. & Ho, K.M. (1996). Genetic-algorithm energy minimization for point charges on a sphere, *Phys. Rev. B*, 53(4), pp. 1740-1743, ISSN 0163-1829

Onsager, L. (1944). Crystal statistics in a two-dimensional model with an order-disorder transition, *Physical Review*, 65(3), Feb. 1944, pp. 117-149

Orlova, G.I. & Dorfman, Y.G. (1972). Finding the maximum cut in a graph, *Engr. Cybernetics*, 10, pp. 502-506

Pang, T. (1997). *An Introduction to Computational Physics*, Cambridge University Press, ISBN 0521825695, New York

Pelikan, M. & Goldberg, D.E. (2003). Hierarchical BOA Solves Ising Spin Glasses and MAXSAT, *Proceedings of GECCO-2003*, pp. 1271-1282, ISBN 3-540-40603-4, Genetic and Evolutionary Computation, Jul. 2003, Springer, Berlin

Pelikan, M. Hartmann, A.K., & Sastry, K. (2006). Hierarchical BOA, Cluster Exact Approximation, and Ising Spin Glasses, *Proceedings of PPSN 2006*, pp. 122-131, ISBN 978-3-540-38990-3, Parallel Problem Solving from Nature - IX, Springer, Berlin

Perez-Garrido, A., Ortuno, M., Cuevas, E. & Ruiz, J. (1996). Many-particle jumps algorithm and Thomson's problem, *Journal of Physics A: Mathematical and General*, 29(9), May 1996, pp. 1973-1978, ISSN 0305-4470

Smale, S. (2000), Mathematical problems for the next century, In: *Mathematics: frontiers and perspectives*, V.I. Arnold, M. Atiyah, P. Lax, B. Mazur (Ed.), pp. 271-294, American Mathematical Society, ISBN 0821820702, Providence, USA

Travesset, A. (2005). Ground state of a large number of particles on a frozen topography, *Physical Review E*, 72(3), September, 2005, pp. 036110, ISSN 0031-9007

5

Particle Swarm Optimization Algorithm for Transportation Problems

Han Huang[1] and Zhifeng Hao[2]
[1]*School of Software Engineering, South China University of Technology*
[2]*College of Mathematical Science, South China University of Technology*
P. R. China

1. Brief Introduction of PSO

Particle swarm optimization (PSO) is a newer evolutionary computational method than genetic algorithm and evolutionary programming. PSO has some common properties of evolutionary computation like randomly searching, iteration time and so on. However, there are no crossover and mutation operators in the classical PSO. PSO simulates the social behavior of birds: Individual birds exchange information about their position, velocity and fitness, and the behavior of the flock is then influenced to increase the probability of migration to regions of high fitness. The framework of PSO can be described as Figure 1.

1. Initialize K Particles $X_1, X_2, ..., X_K$, calculating $pbest_1, ..., pbest_K$ and $gbest, t=1$;
2. For $i=1,...,K$ and $j=1,...,N$, update particles and use X_i to refresh $pbest_i$ and $gbest$; (shown as equation 1 and 2)
3. $t=t+1$; If $t > max\_gen$, output $gbest$ and exit; else, return Step 2.

Figure 1. The framework of classical PSO

In the optimal size and shape design problem, the position of each bird is designed as variables x, while the velocity of each bird v influences the incremental change in the position of each bird. For particle d Kennedy proposed that position x^d be updated as:

$$x^d_{t+1} = x^d_t + v^d_{t+1} \tag{1}$$

$$v^d_{t+1} = v^d_t + c_1 r_1 (p^d_t - x^d_t) + c_2 r_2 (p^g_t - x^d_t) \tag{2}$$

Here, p^d_t is the best previous position of particle d at time t, while p^g_t is the global best position in the swarm at time t. r_1 and r_2 are uniform random numbers between 0 and 1, and $c_1 = c_2 = 2$.

2. Particle Swarm Optimization for linear Transportation Problem

2.1 Linear and Balance Transportation Problem

The transportation problem (TP) is one of the fundamental problems of network flow optimization. A surprisingly large number of real-life applications can be formulated as a TP. It seeks determination of a minimum cost transportation plan for a single commodity from a number of sources to a number of destinations. So the LTP can be described as: Given there are n sources and m destinations. The amount of supply at source i is a_i and the demand at destination j is b_j. The unit transportation cost between source i and destination j is c_{ij}. x_{ij} is the transport amount from source i to destination j, and the LTP model is:

$$\begin{aligned}
&\min \quad z=\sum_{i=1}^{n}\sum_{j=1}^{m} c_{ij}x_{ij} \\
&\text{s.t.} \quad \sum_{j=1}^{m} x_{ij} \le a_i \qquad i=1,2,\ldots,n \\
&\sum_{i=1}^{n} x_{ij} \ge b_j \qquad j=1,2,\ldots,m \cdot \\
&x_{ij} \ge 0; \qquad i=1,2,\ldots,n;\ j=1,2,\ldots,m
\end{aligned} \tag{3}$$

TP has been paid much attention to and classified into several types of transmutation. According to the nature of object function, there are four types: (1) linear TP and nonlinear TP. (2) single objective TP and multi objective. Based on the type of constraints, there are planar TP and solid TP. The single object LTP dealt with in this paper is the basic model for other kinds of transportation problems.

A special LTP called balanced LTP is considered as follows:

$$\begin{aligned}
&\min \quad z=\sum_{i=1}^{n}\sum_{j=1}^{m} c_{ij}x_{ij} \\
&\text{s.t.} \quad \sum_{j=1}^{m} x_{ij} = a_i \qquad i=1,2,\ldots,n \\
&\sum_{i=1}^{n} x_{ij} = b_j \qquad j=1,2,\ldots,m . \\
&\sum_{i=1}^{n} a_i = \sum_{j=1}^{m} b_j \\
&x_{ij} \ge 0; \qquad i=1,2,\ldots,n;\ j=1,2,\ldots,m
\end{aligned} \tag{4}$$

The fact that a $n \times m$ LTP can be changed into a $n \times (m+1)$ balanced LTP, can be found in operational research because the demand at destination $m+1$ could be calculated by $b_{m+1} = \sum_{i=1}^{n} a_i - \sum_{j=1}^{m} b_j$ with the condition $\sum_{i=1}^{n} a_i \geq \sum_{j=1}^{m} b_j$.

2.2 Initialization of PSO for Linear and Balance Particle Swarm Optimization

A particle $X = \begin{bmatrix} x_{11} & \ldots & x_{1m} \\ \ldots & \ldots & \ldots \\ x_{n1} & \ldots & x_{nm} \end{bmatrix}$ stands for a solution to LTP. There are nm particles initialized to form nm initial solutions in the initialization. Every element of set N can be chosen as the first assignment to generate the solutions dispersedly, which is good for obtaining the optimal solution in the iteration.

If a LTP is balanced, the following procedure can be used to obtain an initial solution:

```
program GetOnePrimal (var X: Particle, first: int)
  var   i,j,k: integer;
        N: Set of Integer;

  begin
    k :=0;
      N := {1,2,…,nm};

      repeat
          if k=0 then
              k:=first ;

          else
              k:= a random element in N;
          i := ⌊(k-1)/m⌋+1;
          j := ((k-1) mod m) +1;
          xij := min {ai, bj};
          ai := ai - xij;
          bi := bi - xij;
          N := N \ {k};
      until N is empty
    end.
```

And the initialization of PSO-TP can be designed as follows:

```
program Initialization
  var   i : integer;
  begin
   Get a balanced LTP;
         i := 1;
         repeat
            GetOnePrimal (Xi, i);
            i := i+1;
         until i > nm
  end.
```

2.3 Updating Rule of PSO for Linear and Balance Particle Swarm Optimization

In PSO algorithm, a new solution can be obtained by using the position updating rule as equations 2 and 3. However, the classical rule is unable to meet such constraints of TP as $\sum_{j=1}^{m} x_{ij} = a_i$ and $\sum_{i=1}^{n} x_{ij} = b_j$. A new rule is designed to overcome this shortcoming. For particle d, we propose that position X^d $(n \times m)$ be updated as

$$V^d{}_{t+1} = \begin{cases} \varphi_1(P^d{}_t - X^d{}_t) + \varphi_2(P^g{}_t - X^d{}_t) & t = 0 \\ \lambda_1 V^d{}_t + \lambda_2[\varphi_1(P^d{}_t - X^d{}_t) + \varphi_2(P^g{}_t - X^d{}_t)] & t > 0 \end{cases} \tag{5}$$

$$X^d{}_{t+1} = V^d{}_{t+1} + X^d{}_t \tag{6}$$

where $P^g{}_t \neq X^d{}_t$ and $P^d{}_t \neq X^d{}_t$.

If $P^g{}_t = X^d{}_t$ and $P^d{}_t \neq X^d{}_t$, $\varphi_1 = 1$. If $P^g{}_t \neq X^d{}_t$ and $P^d{}_t = X^d{}_t$, $\varphi_2 = 1$. If $P^g{}_t = X^d{}_t$ and $P^d{}_t = X^d{}_t$, $\lambda_1 = 1$.

$P^d{}_t(n \times m)$ is the best previous position of particle d at time t, while $P^g{}_t(n \times m)$ is the global best position in the swarm at time t. φ_1 and φ_2 are uniform random numbers in (0, 1), meeting $\varphi_1 + \varphi_2 = 1$, while λ_1 is a uniform random number between [0.8, 1.0) and $\lambda_2 = 1 - \lambda_1$.

$$if \quad t = 0,\ X^d{}_{t+1} = V^d{}_{t+1} + X^d{}_t$$

$$= \varphi_1(P^d{}_t - X^d{}_t) + \varphi_2(P^g{}_t - X^d{}_t) + X^d{}_t = (\varphi_1 P^d{}_t + \varphi_2 P^g{}_t) - (\varphi_1 X^d{}_t + \varphi_2 X^d{}_t) + X^d{}_t$$

$$\sum_{j=1}^{m} x^{d}{}_{ij}(t+1)$$

$$= \varphi_1 \sum_{j=1}^{m} p^{d}{}_{ij}(t) + \varphi_2 \sum_{j=1}^{m} p^{g}{}_{ij}(t) - (\varphi_1 \sum_{j=1}^{m} x^{d}{}_{ij}(t) + \varphi_2 \sum_{j=1}^{m} x^{d}{}_{ij}(t)) + \sum_{j=1}^{m} x^{d}{}_{ij}(t)$$

$$= \varphi_1 a_i + \varphi_2 a_i - (\varphi_1 a_i + \varphi_2 a_i) + a_i = a_i$$

$$\sum_{i=1}^{n} x^{d}{}_{ij}(t+1)$$

$$= \varphi_1 \sum_{i=1}^{n} p^{d}{}_{ij}(t) + \varphi_2 \sum_{i=1}^{n} p^{g}{}_{ij}(t) - (\varphi_1 \sum_{i=1}^{n} x^{d}{}_{ij}(t) + \varphi_2 \sum_{i=1}^{n} x^{d}{}_{ij}(t)) + \sum_{i=1}^{n} x^{d}{}_{ij}(t)$$

$$= \varphi_1 b_j + \varphi_2 b_j - (\varphi_1 b_j + \varphi_2 b_j) + b_j = b_j$$

$$if \quad t > 0,$$

$$X^{d}{}_{t+1} = V^{d}{}_{t+1} + X^{d}{}_{t}$$

$$= \lambda_1 V^{d}{}_{t} + \lambda_2 [\varphi_1 (P^{d}{}_{t} - X^{d}{}_{t}) + \varphi_2 (P^{g}{}_{t} - X^{d}{}_{t})] + X^{d}{}_{t}$$

$$= \lambda_1 V^{d}{}_{t} + \lambda_2 [(\varphi_1 P^{d}{}_{t} + \varphi_2 P^{g}{}_{t}) - (\varphi_1 X^{d}{}_{t} + \varphi_2 X^{d}{}_{t})] + X^{d}{}_{t}$$

$$= \lambda_1 (X^{d}{}_{t} - X^{d}{}_{t-1}) + \lambda_2 [(\varphi_1 P^{d}{}_{t} + \varphi_2 P^{g}{}_{t}) - (\varphi_1 X^{d}{}_{t} + \varphi_2 X^{d}{}_{t})] + X^{d}{}_{t}$$

$$\sum_{j=1}^{m} x^{d}{}_{ij}(t+1) = \lambda_1 (\sum_{j=1}^{m} x^{d}{}_{ij}(t) - \sum_{j=1}^{m} x^{d}{}_{ij}(t-1)) + \lambda_2 (\varphi_1 a_i + \varphi_2 a_i - (\varphi_1 a_i + \varphi_2 a_i)) + a_i$$

$$= \lambda_1 (a_i - a_i) + 0 + a_i = a_i$$

$$\sum_{i=1}^{n} x^{d}{}_{ij}(t+1)$$

$$= \lambda_1 (\sum_{i=1}^{n} x^{d}{}_{ij}(t) - \sum_{i=1}^{n} x^{d}{}_{ij}(t-1)) + \lambda_2 (\varphi_1 b_j + \varphi_2 b_j - (\varphi_1 b_j + \varphi_2 b_j)) + b_j$$

$$= \lambda_1 (b_j - b_j) + 0 + b_j = b_j$$

Therefore, X^d_{t+1} would meet the condition that $\sum_{j=1}^{m} x^d_{ij}(t+1) = a_i$ and $\sum_{i=1}^{n} x^d_{ij}(t+1) = b_j$ with the function of Formulae 5 and 6. However, the new rule cannot ensure the last constraint that $x_{ij} \geq 0, i = 1,..,n, j = 1,...,m$. In the following section, an extra operator is given to improve the algorithm.

2.4 Negative Repair Operator

A particle of PSO-TP (Formula 7) will be influenced by the negative repair operator if $x_{ki} < 0, k = 1,...,n, i = 1,...,m$, which is indicated as follows:

$$X = \begin{bmatrix} x_{11} & \cdots & x_{1i} & \cdots & x_{1m} \\ \cdots & \cdots & \cdots & \cdots & \cdots \\ x_{k1} & \cdots & x_{ki} & \cdots & x_{km} \\ \cdots & \cdots & \cdots & \cdots & \cdots \\ x_{l1} & \cdots & x_{li} & \cdots & x_{lm} \\ \cdots & \cdots & \cdots & \cdots & \cdots \\ x_{n1} & \cdots & x_{ni} & \cdots & x_{nm} \end{bmatrix} \tag{7}$$

---

program RepairOnePos (var X: Particle, k,i: int)
begin

select the maximum element signed as x_{li} in Col. i;

$x_0 := x_{ki}$, $x_{li} := x_{li} - |x_0|$, $x_{ki} := 0$;

change elements in Row.k into $x_{kj} := \begin{cases} x_{kj} & x_{kj} = 0 \\ x_{kj} - \dfrac{|x_0|}{u} & x_{kj} > 0 \end{cases}$;

(u is the number of times when the following condition $x_{kj} > 0, j = 1,...,m$ is met)

change elements in Row. l into $x_{lj} := \begin{cases} x_{lj} & x_{kj} = 0 \\ x_{lj} + \dfrac{|x_0|}{u} & x_{kj} > 0 \end{cases}$;

end.

As a result, the procedure of negative repair operator can be described as:

```
program NegativeRepair (var X: Particle)
var   i,j: integer;
begin
   if some element of X is negative then
        repeat
          If x_ij<0 is found then
                  RepairOnePos (X, i, j);
        until Every element of X is not negative
end.
```

2.5 PSO Mutation

Mutation is a popular operator in Genetic Algorithm, and a special PSO mutation is designed to help PSO-TP change the partial structure of some particles in order to get new types of solution. PSO-TP cannot fall into the local convergence easily because the mutation operator can explore the new solution.

```
program PSOMutation (var X: Particle)
begin
  Obtain p and q randomly meeting 0<p<n and 0<q<m;
   Select p rows {i1,...ip} and q lines {j1,...jq} randomly from matrix X to form a small matrix
Y (yij,i=1,...,p,j=1,...,q);
```

$$a^y_i = \sum_{j \in \{j_1,\ldots,j_q\}} x_{ij} \qquad (i = i_1,\ldots,i_p)$$

$$b^y_j = \sum_{i \in \{i_1,\ldots,i_p\}} x_{ij} \qquad (j = j_1,\ldots,j_q)$$

```
   Use a method like the one in initialization to form the initial assignment for Y;
  Update X with Y;
end.
```

2.6 The Structure of PSO-TP

According to the setting above, the structure of PSO-TP is shown as:

```
program PSO-TP (problem: balanced LTP of n×m size, pm: float)
var t:integer;
begin
   t:=0;
   Initialization;
```

Obtain $P^g{}_0(n \times m)$ and $P^d{}_0(n \times m)$(d=1,…,n×m);

repeat

t:=t+1;

Calculate $X^d{}_t$ with Formula 5 and 6 (d=1,…,n×m);

NegativeRepair($X^d{}_t$)(d=1,…,n×m);

Carry out PSOMutation($X^d{}_t$) by the probability pm;

Update $P^g{}_t(n \times m)$ and $P^d{}_t(n \times m)$(d=1,…,n×m);

until meeting the condition to stop

end.

3. Numerical Results

There are two experiments in this section: one is comparing PSO-TP with genetic algorithm (GA) in some integer instances and the second is testing the performance of PSO-TP in the open problems. Both of the experiments are done at a PC with 3.06G Hz, 512M DDR memory and Windows XP operating system. GA and PSO-TP would stop when no better solution could be found in 500 iterations, which is considered as a virtual convergence of the algorithms. The probability of mutation in PSO-TP is set to be 0.05.

Problem\ five runs	PSO-TP Min	PSO-TP Ave	GA Min	GA Ave	PSO-TP Time(s)	GA Time(s)
P1 (3*4)	152	152	152	153	0.015	1.72
P2 (4*8)	287	288	290	301	0.368	5.831
P3 (3*4)	375	375	375	375	0.028	0.265
P4 (3*4)	119	119	119	119	0.018	1.273
P5 (3*4)	85	85	85	85	0.159	0.968
P6*(15*20)	596	598	-	-	36.4	-

Table 1. Comparison Between PSO-TP and GA

As Table 1 shows, both the minimum cost and average cost obtained by PSO-TP are less than those of GA. Furthermore, the time cost of PSO-TP is much less than that of GA. In order to verify the effectiveness of PSO-TP, 9 real number instances are computed and the results are shown in Table 2. Since GA is unable to deal with the real number LTP directly, only PSO-T is tested.

Problem\five runs	Optimal Value	PSO-TP Average	PSO-TP Time(s)
No.1	67.98	67.98	0.02
No.2	1020	1020	0.184
No.4	13610	13610	0.168
No.5	1580	1580	0.015
No.6	98	98	0.023
No.7	2000	2000	0.015
No.8	250	250	<0.001
No.9	215	215	0.003
No.10	110	110	0.012

Table 2. Performance of PSO-TP in open problems

According to the results in Table 2, PSO-TP can solve the test problems very quickly. The efficiency of PSO-TP may be due to the characteristic of PSO algorithm and the special operators. Through the function of the new position updating rule and negative repair operator, the idea of PSO is introduced to solve LTP successfully. The nature of PSO can accelerate the searching of the novel algorithm, which would also enable PSO-TP to get the local best solution. What's more, the PSO mutation as an extra operator can help PSO-TP to avoid finishing searching prematurely. Therefore, PSO-TP can be a novel effective algorithm for solving TP.

4. Particle Swarm Optimization for Non-linear Transportation Problem

4.1 Non-linear and Balance Transportation Problem

The unit transportation cost between source i and destination j is $f_{ij}(x_{ij})$ where x_{ij} is the transportation amount from source i to destination j, and TP model is:

$$\min \quad z=\sum_{i=1}^{n}\sum_{j=1}^{m} f_{ij}(x_{ij})$$

$$\text{s.t.} \quad \sum_{j=1}^{m} x_{ij} \le a_i \qquad i=1,2,\ldots,n \tag{8}$$

$$\sum_{i=1}^{n} x_{ij} \ge b_j \qquad j=1,2,\ldots,m \cdot$$

$$x_{ij} \ge 0; \qquad i=1,2,\ldots,n; j=1,2,\ldots,m$$

According to the nature of object function, there are four types: linear TP in which the function $f_{ij}(x_{ij})$ is linear and nonlinear TP in which $f_{ij}(x_{ij})$ is non-linear, as well as single objective and multi-objective TP. Based on the types of constraints, there are planar TP and solid TP. The single object NLTP is dealt with in this paper. In many fields like railway transportation, the relation between transportation amount and price is often non-linear, so NLTP is an important for application.

4.2 Framework of PSO for Non-linear TP

In the population of PSO-NLTP, an individual $X_i = \begin{bmatrix} x_{11} & \cdots & x_{1m} \\ \cdots & \cdots & \cdots \\ x_{n1} & \cdots & x_{nm} \end{bmatrix}$ stands for a solution to NLTP (Exp. 2), where $n \times m$ is the population size. There are $n \times m$ individuals initialized to form $n \times m$ initial solutions in the initialization. The initialization and mutation are the same as the ones in PSO-LTP (Section 2.2 and 2.5).
And the framework of PSO-NLTP is given:

Algorithm: PSO-NLTP
Input: NLTP problem (Exp. 8)
begin
  Initialization;
  Setting parameters;
 repeat
  Updating rule;
  Mutation;
  Updating the current optimal solution
 until meeting the condition to stop
end.
Output: Optimal solution for NLTP

In the parameter setting, The parameters of PSO-NLTP are all set adaptively: as the population size is $n \times m$, the size of mutation matrix Y is set randomly meeting 0<p<n and 0<q<m and the mutation probability P_m is calculated by $P_m = 0.005 \times N_t$, where $N_t = 1$ when $X_{best}^{(t)}$ is updated and $N_t = N_t + 1$ when $X_{best}^{(t)}$ remains the same as $X_{best}^{(t-1)}$.

4.3 Updating Rule of PSO-NLTP

As one of the important evolutionary operator, recombination is designed to optimize the individuals and make them meet the constraints of supply and demand as $\sum_{j=1}^{m} x_{ij} = a_i$ and

$\sum_{i=1}^{n} x_{ij} = b_j$ (Exp. 4). At the beginning of an iteration, every individual is recombined by the following expression.

$$X_i^{(t+1)} = \varphi_1 X_i^{(t)} + \varphi_2 X_{best}^{(t)} + \varphi_3 X_{random}^{(t)} \tag{9}$$

$X_{best}^{(t)}$ is the best particle found by PSO-NLTP form iteration 0 to t. $X_{random}^{(t)}$ is the particle formed randomly (by sub-algorithm GetOnePrimal in section 2.2) for the updating rule of $X_i^{(t)}$. φ_1, φ_2 and φ_3 are the weight terms meeting $\varphi_1 + \varphi_2 + \varphi_3 = 1$, which are calculated as Exp 4-6 show, where $f(X_i^{(t)})$ is the cost of the solution for TP (Exp. 4).

$$\varphi_1 = f(X_i^{(t)})^{-1} / \left(f(X_i^{(t)})^{-1} + f(X_{best}^{(t)})^{-1} + f(X_{random}^{(t)})^{-1} \right) \tag{10}$$

$$\varphi_2 = f(X_{best}^{(t)})^{-1} / \left(f(X_i^{(t)})^{-1} + f(X_{best}^{(t)})^{-1} + f(X_{random}^{(t)})^{-1} \right) \tag{11}$$

$$\varphi_3 = f(X_{random}^{(t)})^{-1} / \left(f(X_i^{(t)})^{-1} + f(X_{best}^{(t)})^{-1} + f(X_{random}^{(t)})^{-1} \right) \tag{12}$$

$X_i^{(t+1)}$ can be considered as a combination of $X_i^{(t)}$, $X_{best}^{(t)}$ and $X_{random}^{(t)}$ based on the their quality, and proved to meet the constraints of supply and demand.

$$\sum_{j=1}^{m} x_{ij}^{(t+1)} = \sum_{j=1}^{m} (\varphi_1 x_{ij}^{(t)} + \varphi_2 x_{i,j,best}^{(t)} + \varphi_3 x_{i,j,random}^{(t)})$$

$$= \varphi_1 \sum_{j=1}^{m} x_{ij}^{(t)} + \varphi_2 \sum_{j=1}^{m} x_{i,j,best}^{(t)} + \varphi_3 \sum_{j=1}^{m} x_{i,j,random}^{(t)}$$

$$= \varphi_1 a_i + \varphi_2 a_i + \varphi_3 a_i = a_i \quad (i = 1, \ldots, n)$$

$$\sum_{i=1}^{n} x_{ij}^{(t+1)} = \sum_{i=1}^{n} (\varphi_1 x_{ij}^{(t)} + \varphi_2 x_{i,j,best}^{(t)} + \varphi_3 x_{i,j,random}^{(t)})$$

$$= \varphi_1 \sum_{i=1}^{n} x_{ij}^{(t)} + \varphi_2 \sum_{i=1}^{n} x_{i,j,best}^{(t)} + \varphi_3 \sum_{i=1}^{n} x_{i,j,random}^{(t)}$$

$$= \varphi_1 b_j + \varphi_2 b_j + \varphi_3 b_j = b_j \quad (j = 1, \ldots, m)$$

Furthermore, the recombination rule can also ensure the positive constraint that $x_{ij}^{(t+1)} = \varphi_1 x_{ij}^{(t)} + \varphi_2 x_{i,j,best}^{(t)} + \varphi_3 x_{i,j,random}^{(t)} \geq 0, i = 1,..,n, j = 1,...,m$.

4.4 Numerical Results

There are 56 NLTP instances computed in the experiment, of which the results are shown in this section. The experiment is done at a PC with 3.06G Hz, 512M DDR memory and Windows XP operating system. The NLTP instances are generated by replacing the linear cost functions of the open problems with the non-linear functions. The methods which are effective for linear TP cannot deal with NLTP for the complexity of non-linear object function. The common NLTP cost functions are indicated in Table 1.

Problem	Transportation Cost Functions
No.1	$f_{ij}(x_{ij}) = c_{ij} x_{ij}^2$
No.2	$f_{ij}(x_{ij}) = c_{ij} \sqrt{x_{ij}}$
No.3	$f_{ij}(x_{ij}) = \begin{cases} c_{ij}(\frac{x_{ij}}{S}), & if\ 0 \leq x_{ij} < S \\ c_{ij}, & if\ S < x_{ij} \leq 2S \\ c_{ij}(1 + \frac{x_{ij} - 2S}{S}), & if\ 2S < x_{ij} \end{cases}$
No.4	$f_{ij}(x_{ij}) = c_{ij} x_{ij} [\sin(x_{ij} \frac{5\pi}{4S}) + 1]$

Table 3. NLTP cost functions [15]

The comparison between PSO-NLTP and EP with penalty strategy only indicates whether the recombination of PSO-NLTP is better at dealing with the constraints of NLTP (Exp. 8) than penalty strategy of EP. There cannot be any conclusion that PSO-NLTP or EP is better than the other because they are the algorithms for different applications. The three algorithms are computed in 50 runs independently, and the results are in Table 4 and Table 5. They would stop when no better solution could be found in 100 iterations, which is considered as a virtual convergence of the algorithms.

NLTP instances in Table 4 are formed with the non-linear functions (shown in Table 3) and the problems. And the instances in Table 5 are formed with the non-linear functions and the problems. We set $S = \sum_{i=1}^{n} a_i / 10$ in function No.3 and $S = 1$ in function No.4 in the experiment.

Problem	PSO-NLTP Average	GA Average	EP Average	PSO-NLTP Time(s)	GA Time(s)	EP Time(s)
No.1-1	8.03	8.10	8.36	0.093	0.89	0.109
No.1-2	112.29	114.25	120.61	0.11	0.312	0.125
No.1-4	1348.3	1350.8	1476.1	0.062	0.109	0.078
No.1-5	205.9	206.3	216.1	0.043	0.125	0.052
No.1-6	12.64	12.72	13.53	0.062	0.75	0.078
No.1-7	246.9	247.6	256.9	0.088	0.32	0.093
No.1-8	84.72	84.72	87.5	<0.001	0.015	<0.001
No.1-9	44.64	44.65	46.2	<0.001	0.046	<0.001
No.1-10	24.85	24.97	25.83	<0.001	0.032	<0.001
No.2-1	155.3	155.3	168.5	<0.001	0.016	<0.001
No.2-2	2281.5	2281.5	2696.2	<0.001	0.015	<0.001
No.2-4	28021	28021	30020.2	<0.001	0.015	<0.001
No.2-5	3519.3	3520.4	3583.1	<0.001	0.015	<0.001
No.2-6	264.9	266.5	314.4	<0.001	0.015	<0.001
No.2-7	4576.9	4584.5	5326.0	0.009	0.052	0.012
No.2-8	432.8	432.8	432.8	<0.001	0.015	<0.001
No.2-9	386.3	386.3	386.3	<0.001	0.031	<0.001
No.2-10	195.3	195.3	226.0	<0.001	0.006	<0.001
No.3-1	309.9	310.0	346.6	<0.001	0.093	0.001
No.3-2	4649.2	4650	5415.2	<0.001	0.921	0.012
No.3-4	65496.7	66123.3	68223.3	<0.001	0.105	<0.001
No.3-5	7038.1	7066.6	7220.9	<0.001	1.015	0.001
No.3-6	540	540	672.5	0.001	0.062	0.002
No.3-7	9171.0	9173.2	9833.3	<0.001	0.312	<0.001
No.3-8	1033.4	1033.4	1066.7	<0.001	0.012	<0.001
No.3-9	933.3	933.4	1006.4	0.002	0.147	0.015
No.3-10	480	480	480	0.016	0.046	0.004
No.4-1	107.6	107.8	118.2	0.063	0.159	0.078
No.4-2	1583.5	1585.2	1622	0.062	0.285	0.093
No.4-4	19528.4	19531.3	20119	0.075	0.968	0.068
No.4-5	2466.9	2468.2	2880.2	0.072	0.625	0.046
No.4-6	151.7	152.1	161.9	0.093	1.046	0.167
No.4-7	3171.1	3173.8	3227.5	0.047	0.692	0.073
No.4-8	467.1	467.1	467.1	<0.001	0.036	<0.001
No.4-9	376.3	376.3	382.5	<0.001	0.081	0.003
No.4-10	205.9	205.9	227.6	0.026	0.422	0.031

Table 4. Comparison I between PSO-NLTP, GA and EP with penalty strategy

Problem	PSO-NLTP Average	GA Average	EP Average	PSO-NLTP Time(s)	GA Time(s)	EP Time(s)
No.1-11	1113.4	1143.09	1158.2	0.031	0.065	0.046
No.1-12	429.3	440.3	488.3	0.187	1.312	0.203
No.1-13	740.5	740.5	863.6	0.09	2.406	0.781
No.1-14	2519.4	2529.0	2630.3	0.015	0.067	0.016
No.1-15	297.2	297.9	309.2	0.046	0.178	0.058
No.1-16	219.92	220.8	234.6	0.040	1.75	0.060
No.2-11	49.7	51.9	64.2	<0.001	0.001	<0.001
No.2-12	78.4	78.4	104.5	0.001	0.025	<0.001
No.2-13	150.2	150.4	177.9	<0.001	0.015	<0.001
No.2-14	118.6	118.2	148.4	<0.001	0.001	<0.001
No.2-15	64.5	64.5	64.5	<0.001	0.031	<0.001
No.2-16	47.1	47.8	53.4	<0.001	0.015	<0.001
No.3-11	13.3	13.3	13.3	0.015	0.734	0.031
No.3-12	21.0	21.0	26.3	0.018	0.308	0.036
No.3-13	37.2	37.4	43.5	0.171	1.906	0.156
No.3-14	37.5	37.8	46.7	0.011	0.578	0.008
No.3-15	28.3	28.1	33	0.009	0.325	0.013
No.3-16	22.5	23.0	29.6	<0.001	0.059	0.015
No.4-11	8.6	8.8	37.4	0.001	0.106	0.001
No.4-12	20.0	23.1	40.8	0.253	2.328	0.234
No.4-13	49.0	52.3	72.1	0.109	2.031	0.359
No.4-14	47.7	51.2	82.2	0.003	0.629	0.006
No.4-15	11.97	12.06	36.58	0.019	0.484	0.026
No.4-16	2.92	3.08	8.1	0.031	0.921	0.045

Table 5. Comparison II between PSO-NLTP, GA and EP with penalty strategy

As Table 4 and Table 5 indicate, PSO-NLTP performs the best of three in the items of average transportation cost and average computational cost. The NLTP solutions found by EP with penalty strategy cost more than PSO-NLTP and GA, which indicates recombination of PSO-NLTP and crossover of GA handle the constraints of NLTP (Exp. 4) better than the penalty strategy. However, EP with penalty strategy cost less time than GA to converge because the crossover and mutation operator of GA is more complicated. PSO-NLTP can cost the least to obtain the best NLTP solution of the three tested methods. Its recombination makes the particles feasible and evolutionary for optimization. The combination of updating rule and mutation operators can play a part of global searching quickly, which makes PSO-NLTP effective for solving NLTPs.

5. Discussions and Conclusions

Most of the methods that solve linear transportation problems well cannot handle the nonlinear TP. An particle swarm optimization algorithm named PSO-NLTP is proposed in the present paper to deal with NLTP. The updating rule of PSO-NLTP can make the particles of the swarm optimally in the feasible solution space, which satisfies the constraints of NLTP. A mutation operator is added to strengthen the global optimal capacity of PSO-NLTP. In the experiment of computing 56 NLTP instances, PSO-NLTP performs much better than GA and EP with penalty strategy. All of the parameters of PSO-NLTP are set adaptively in the iteration so that it is good for the application of the proposed algorithm. Moreover, PSO-NLTP can also solve linear TPs.

The design of the updating rule of PSO can be considered as an example for solving optimization problems with special constraints. The operator is different from other methods such as stochastic approach, greedy decoders and repair mechanisms, which are to restrict the searching only to some feasible sub-space satisfying the constraints. It uses both the local and global heuristic information for searching in the whole feasible solution space. Furthermore, through the initial experimental result, it performs better than the penalty strategy which is another popular approach for handling constraints.

6. References

Papamanthou C., Paparrizos K., and Samaras N., Computational experience with exterior point algorithms for the transportation problem, *Applied Mathematics and Computation*, vol. 158, pp. 459-475, 2004. [1]

Vignaux G.A. and Michalewicz Z., A genetic algorithm for the linear transportation problem, *IEEE Transactions on Systems, Man, and Cybernetics*, vol. 21, no. 2, MARCWAPRIL, pp.445-452, 2004. [2]

Hitchcock F., The distribution of a product from several sources to numerous location, *Journal of Mathematical Physics*, vol. 20, pp. 224-230, 1941. [3]

Michalewicz Z., et al, A non-Standard Genetic Algorithm for the Nonlinear Transportation Problems, *ORSA Journal on Computing*, vol. 3, no. 4, pp.307-316, 1991. [4]

Li Y.Z., Ida K.C. and Gen M., Improved genetic algorithm for solving multi objective solid transportation problem with fuzzy numbers, *Computers ind. Engng*, vol. 33, no.3-4, pp. 589-592, 1997. [5]

Gen M., et al, Solving bicriteria solid transportation problem by genetic algorithms, *Proceedings of the 16th International Conference on computers and industrial engineering*, Ashikaga, Japan, pp.572-575, 1994. [6]

Dantzig G.B., Application of the simplex method to a transportation problem, in: T.C. Koopmans (Ed.), *Activity of production and application*, John Wiley & Sons, NY, pp. 359-373, 1951. [7]

Orlin J.B., Plotkin S.A. and Tardos E., Polynomial dual network simplex algorithms, *Math. Program*, vol. 60, pp. 255-276, 1993. [8]

Paparrizos K., An exterior point simplex algorithm for general linear problems, *Ann. Oper. Res.*, vol. 32, pp. 497-508, 1993. [9]

Papamanthou C., Paparrizos K. and Samaras N., Computational experience with exterior point algorithms for the transportation problem, *Applied Mathematics and Computation*, vol. 158, pp. 459-475, 2004. [10]

Sharma R.R.K. and Sharma K.D., A new dual based procedure for the transportation problem, *European Journal of Operational Research*, vol. 122, pp. 611-624, 2000. [11]

Yang X. and Gen M., Evolution program for bicriteria transportation problem, *Proceedings of the 16th International Conference on computers and industrial engineering*, Ashikaga, Japan, pp. 451-454, 1994. [12]

Kennedy J. and Eberhart R.C., Particle swarm optimization, in *Proc IEEE Int. Conf. Neural Networks*, Perth, Austalia,pp. 1942-1948, Nov. 1995. [13]

Kennedy J., The particle swarm: Social adaptation knowledge, in *Proc. 1997 Int. Conf. Evolutionary Computation*, Indianapolis, IN, pp. 303-308, Apr. 1997. [14]

Fourie P.C. and Groenwold A.A., The particle swarm optimization algorithm in size and shape optimization, *Struct Multidisc Optim*, vol. 23, pp. 259-267, 2000. [15]

Shi Y.H. and Eberhart R.C., A modified particle swarm optimizer, *Proc. Int. Conf. On Evolutionary Computation*, pp.69-73, 1998. [16]

6

Particle Swarm Optimization and Other Metaheuristic Methods in Hybrid Flow Shop Scheduling Problem

M. Fikret Ercan
Singapore Polytechnic School of Electrical and Electronic Engineering
Singapore

1. Introduction

Multiprocessor task scheduling is a generalized form of classical machine scheduling where a task is processed by more than one processor. It is a challenging problem encountered in wide range of applications and it is vastly studied in the scheduling literature (see for instance (Chan & Lee, 1999 and Drozdowski, 1996) for a comprehensive introduction on this topic). However, Drozdowski (1996) shows that multiprocessor task scheduling is difficult to solve even in its simplest form. Hence, many heuristic algorithms are presented in literature to tackle multiprocessor task scheduling problem. Jin et al. (2008) present a performance study of such algorithms. However, most of these studies primarily concerned with a single stage setting of the processor environment. There are many practical problems where multiprocessor environment is a flow-shop that is it is made of multiple stages and tasks have to go through one stage to another.

Flow-shop scheduling problem is also vastly studied in scheduling context though most of these studies concerned with single processor at each stage (see for instance Linn & Zhang, 1999, Dauzère-Pérès & Paulli, 1997). With the advances made in technology, in many practical applications, we encounter parallel processors at each stage instead of single processors such as parallel computing, power system simulations, operating system design for parallel computers, traffic control in restricted areas, manufacturing and many others (see for instance (Krawczyk & Kubale, 1985, Lee & Cai, 1999, Ercan & Fung, 2000, Caraffa et. al.,2001)). This particular problem is defined as hybrid flow-show with multiprocessor tasks in scheduling terminology and minimizing the schedule length (makespan) is the typical scheduling problem addressed. However, Brucker & Kramer (1995) show that multiprocessor flow-shop problem to minimize makespan is also NP-hard. Gupta (1988) showed that hybrid flow-shop even with two stages is NP-hard. Furthermore, the complexity of the problem increases with the increasing number of stages.

Multiprocessor task scheduling in a hybrid flow-shop environment has recently gained the attention of the research community. To the best of our knowledge, one of the earliest papers that deal with this problem in the scheduling literature is by Oğuz and Ercan, 1997. However, due to the complexity of the problem, in the early studies (such as (Lee & Cai, 1999, Oğuz et. al., 2003)) researchers targeted two layer flow-shops with multiprocessors. Simple list based heuristics as well as meta-heuristics were introduced for the solution

(Oğuz et al.,2003, Jdrzęjowicz & Jdrzęjowicz,2003, Oğuz et al.,2004). Apparently, a broader form of the problem will have arbitrary number of stages in the flow-shop environment. This is also studied recently and typically metaheuristic algorithms applied to minimize the makespan such as population learning algorithm (Jdrzęjowicz & Jdrzęjowicz, 2003), tabu search (Oğuz et al.,2004), genetic algorithm (Oğuz & Ercan,2005) and ant colony system (Ying & Lin,2006). Minimizing the makespan is not the only scheduling problem tackled; recently Shiau et al. (2008) focused on minimizing the weighted completion time in proportional flow shops.

These metaheuristic algorithms produce impressive results though they are sophisticated and require laborious programming effort. However, of late particle swarm optimization (PSO) is gaining popularity within the research community due to its simplicity. The algorithm is applied to various scheduling problems with notable performance. For instance, Sivanandam et al. (2007) applied PSO to typical task allocation problem in multiprocessor scheduling. Chiang et al. (2006) and Tu et al. (2006) demonstrate application of PSO to well known job shop scheduling problem.

PSO, introduced by Kennedy & Eberhart (1995), is another evolutionary algorithm which mimics the behaviour of flying birds and their communication mechanism to solve optimization problems. It is based on a constructive cooperation between particles instead of survival of the fittest approach used in other evolutionary methods. PSO has many advantages therefore it is worth to study its performance for the scheduling problem presented here. The algorithm is simple, fast and very easy to code. It is not computationally intensive in terms of memory requirements and time. Furthermore, it has a few parameters to tune.

This chapter will present the hybrid flow-shop with multiprocessor tasks scheduling problem and particle swarm optimization algorithm proposed for the solution in details. It will also introduce other well known heuristics which are reported in literature for the solution of this problem. Finally, a performance comparison of these algorithms will be given.

2. Problem definition

The problem considered in this paper is formulated as follows: There is a set J of n independent and simultaneously available jobs where each job is made of Multi-Processor Tasks (*MPT*) to be processed in a multi-stage flow-shop environment, where stage j consists of m_j identical parallel processors ($j=1,2,...,k$). Each $MPT_i \in J$ should be processed on $p_{i,j}$ identical processors simultaneously at stage j without interruption for a period of $t_{i,j}$ $(i=1,2,...,n \text{ and } j=1,2,...,k)$. Hence, each $MPT_i \in J$ is characterized by its processing time, $t_{i,j}$, and its processor requirement, $p_{i,j}$. The scheduling problem is basically finding a sequence of jobs that can be processed on the system in the shortest possible time. The following assumptions are made when modeling the problem:

- All the processors are continuously available from time 0 onwards.
- Each processor can handle no more than one task at a time.
- The processing time and the number of processors required at each stage are known in advance.
- Set-up times and inter-processor communication time are all included in the processing time and it is independent of the job sequence.

3. Algorithms

3.1 The basic PSO algorithm

PSO is initialized with a population of random solutions which is similar in all the evolutionary algorithms. Each individual solution flies in the problem space with a velocity which is adjusted depending on the experiences of the individual and the population. As mentioned earlier, PSO and its hybrids are gaining popularity in solving scheduling problems. A few of these works tackle the flow shop problem (Liu et al.,2005) though application to hybrid flow-shops with multiprocessor tasks is relatively new (Ercan & Fung, 2007, Tseng & Liao, 2008).

In this study, we first employed the global model of the PSO (Ercan & Fung, 2007). In the basic PSO algorithm, particle velocity and position are calculated as follows:

$$V_{id}=W\ V_{id} + C_1R_1(P_{id}-X_{id})+C_2R_2(P_{gd}-X_{id}) \tag{1}$$

$$X_{id}=X_{id}+V_{id} \tag{2}$$

In the above equations, V_{id} is the velocity of particle i and it represents the distance traveled from the current position. W is inertia weight. X_{id} represents particle position. P_{id} is the local best solution (also called as "pbest") and P_{gd} is global best solution (also called as "qutgbest"). C_1 and C_2 are acceleration constants which drive particles towards local and global best positions. R_1 and R_2 are two random numbers within the range of [0, 1]. This is the basic form of the PSO algorithm which follows the following steps:

Algorithm 1: The basic PSO

```
Initialize swarm with random positions and velocities;
begin
repeat
   For each particle evaluate the fitness i.e. makespan of the schedule;
   if current fitness of particle is better than Pid then set Pid to current value;
   if Pid is better than global best then set Pgd to current particle fitness value;
   Change the velocity and position of the particle;
until termination = True
end.
```

The initial swarm and particle velocity are generated randomly. A key issue is to establish a suitable way to encode a shedule (or solution) to PSO particle. We employed the method shown by Xia et al.(2006). Each particle consists of a sequence of job numbers representing the n number of jobs on a machine with k number of stages where each stage has m_j identical processors ($j=1,2,...,k$). The fitness of a particle is then measured with the maximum completion time of all jobs. In our earlier work (Oğuz & Ercan,2005), a list scheduling algorithm is developed to map a given job sequence to the machine environment and to compute the maximum completion time (makespan). A particle with the lowest completion time is a good solution.

Figure 1 shows an example to scheduling done by the list scheduling algorithm. In this example, number of jobs is $n=5$ and a machine is made of two stages $k=2$ where each stage contains four identical processors. Table 1 depicts the list of jobs and their processing times and processor requirements at each stage for this example.

Job #	Stage 1 (j=1)		Stage 2 (j=2)	
i	$p_{i,1}$	$t_{i,1}$	$p_{i,2}$	$t_{i,2}$
1	1	1	2	2
2	3	3	4	2
3	3	3	3	2
4	2	1	2	1
5	1	1	1	1

Table 1. Example jobs and their processing time and processor requirements at each stage

For the schedule shown in Figure 1, it is assumed that a job sequence is given as $S_1 = \{2,3,1,4,5\}$. At stage 1, jobs are iteratively allocated to processors from the list starting from time 0 onwards. As job 2 is the first in the list, it is scheduled at time 0. It is important to note that although there are enough available processors to schedule job 1 at time 0 this will violate the precedence relationship established in the list. Therefore, job 1 is scheduled to time instance 3 together with job 3 and this does not violate the precedence relationship given in S_1 . Once all the jobs are scheduled at first stage, a new list is produced for the succeeding stage based on the completion of jobs at previous stage and the precedence relationships given in S_1. In the new list for stage 2, $S_2 = \{2,1,3,4,5\}$, job 1 is scheduled before job 3 since it is available earlier than job 3. At time instance 7, jobs 3, 4 and 5 are all available to be processed. Job 3 is scheduled first since its completion time is earlier at stage 1. Although, there is enough processor to schedule job 5 at time 8 this will again violate the order given in list S_1 , hence it is scheduled together with job 4. In this particular example, jobs 4 and 5 will be the last to be mapped to stage 2 and the over all completion time of tasks will be 10 units.

The parameters of PSO are set based on our empirical study as well as referring to the experiences of other researchers. The acceleration constants C_1 and C_2 are set to 2.0 and initial population of swarm is set to 100. Inertia weight, W, determines the search behavior of the algorithm. Large values for W facilitate searching new locations whereas small values provide a finer search in the current area. A balance can be established between global and local exploration by decreasing the inertia weight during the execution of the algorithm. This way PSO tends to have more global search ability at the beginning and more local search ability towards the end of the execution. In our PSO algorithm, an exponential function is used to set the inertia weight and it is defined as:

$$W = W_{end} + (W_{start} - W_{end})e^{-\frac{x\alpha}{x_{\max}}} \tag{3}$$

where, W_{start} is the starting, W_{end} is the ending inertia values. W_{start} are W_{end} are set as 1.5 and 0.3 respectively. In addition, x shows the current iteration number and x_{max} shows the maximum iteration number which is set to 10000. An integer constant α is used to manipulate the gradient of the exponentially decreasing W value and it is set to 4.

In this application, X_{id} and V_{id} are used to generate and modify solutions therefore they are rounded off to the nearest integer and limited to a maximum value of n which is the maximum number of jobs. That is position coordinates are translated into job sequence in our algorithm and a move in search space is obtained by modifying the job sequence.

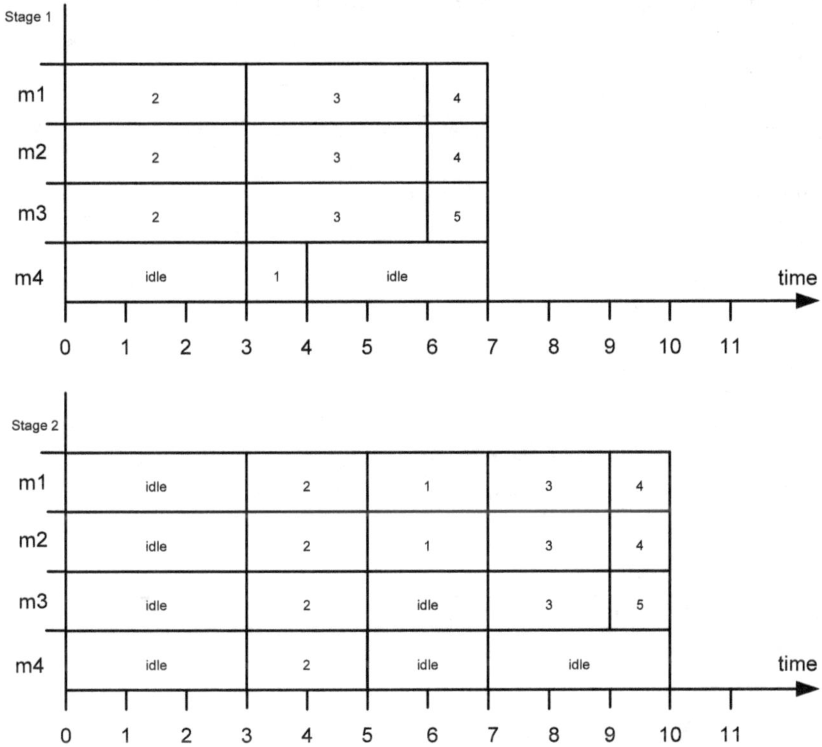

Figure 1. The schedule of job sequence [2, 3, 1, 4, 5] after being allocated to processors of the multilayer system using the list scheduling algorithm. Idle periods of the processors are labeled as idle

3.2 Hybrid PSO algorithms

Although, PSO is very robust and has a well global exploration capability, it has the tendency of being trapped in local minima and slow convergence. In order to improve its performance, many researchers experimented with hybrid PSO algorithms. Poli et al. (2007) give a review on the varations and the hybrids of particle swarm optimisation. Similarly, in scheduling problems, performance of PSO can be improved further by employing hybrid techniques. For instance, Xia & Wu (2006) applied PSO-simulated annealing (SA) hybrid to job shop scheduling problem and test its performance with benchmark problems. Authors conclude that PSO-SA hybrid delivered equal solution quality as compared to other metaheuristic algorithms though PSO-SA offered easier modeling, simplicity and ease of implementation. These findings motivated us to apply PSO and its hybrids to this particular scheduling problem and study its performance.

The basic idea of the hybrid algorithms presented here is simply based on runnign PSO algorithm first and then improving the result by employing a simulated annealing (SA) or tabu search (TS) heuristics. SA and TS introduce a probability to avoid becoming trapped in a local minimum. In addition, by introducing a neighborhood formation and tuning the parameters, it is also possible to enhance the search process. The initial findings of this study are briefly presented in (Ercan,2008). The following pseudo codes show the hybrid algorithms:

Algorithm 2: Hybrid PSO with SA

```
Initialize swarm with random positions and velocities;
   begin
   initialize PSO and SA;
     while (termination !=true)
     do{
      generate swarm;
      compute and find best Pgd;
      }
   set particle that gives best Pgd as initial solution to SA;
      while (Tcurrent>Temp_end)
       do{
       generate neighborhood;
       evaluate and update best solution and temperature;
       }
   end.
```

Algorithm 3: Hybrid PSO with TS

```
Initialize swarm with random positions and velocities;
   begin
   initialize PSO and TS;
      while (termination !=true)
      do{
      generate swarm;
      compute and find best Pgd;
      }
   set particle that gives best Pgd as initial solution to TS;
      while (termination!=true)
       do{
       generate sub set of neighborhoods;
       evaluate and update the best solution;
       update the tabu list;
       }
    end.
```

The initial temperature for PSO-SA hybrid is estimated after 50 randomly permuted neighborhood solutions of the initial solution. A ratio of average increase in the cost to acceptance ratio is used as initial temperature. Temperature is decreased using a simple cooling strategy $T_{current} = \lambda T_{current}\ \text{-}1$. The best value for lambda is experimentally found and set as 0.998. The end temperature is set to 0.01.

A neighbor of the current solution is obtained in various ways.

- Interchange neighborhood: Two randomly chosen jobs from the job list are exchanged.

- Simple switch neighborhood: It is a special case of interchange neighborhood where a randomly chosen job is exchanged with its predecessor.
- Shift neighborhood: A randomly selected job is removed from one position in the priority list and put it into another randomly chosen position.

It is experimentally found that interchange method performs the best amongst all three. The interchange strategy is also found to be the most effective one for generating the sub-neighborhoods for TS.

In the tabu list, a fixed number of last visited solutions are kept. Two methods for updating the tabu list are experimented; elimination of the farthest solution stored in the list, and removing the worst performing solution from the list. In PSO-TS hybrid removing the worst performing solution from the list method is used as it gave a slightly better result.

4. GA algorithm

Genetic algorithms, introduced by Holland (1975), have been widely applied to many scheduling problems in literature (see for instance job-shop environment (Della et al.,1995) and (Dorndorf & Pesch,1995), flow-shop environment (Murata et al., 1996)). Genetic algorithms are also employed in hybrid flow-shops with multiprocessor environment (Oğuz & Ercan, 2005). In this work, authors proposed a new crossover operator, NXO, to be used in the genetic algorithm and compare its performance with well-known PMX crossover. They employed two selection criteria in NXO to minimize the idle time of the processors. Firstly, NXO basically aims to keep the best characteristics of the parents in terms of the neighbouring jobs. That is if two jobs are adjacent to each other in both parents with good fitness values, then NXO tries to keep this structure in the offspring. If there is no such structure, then next criteria is employed in which NXO tries to choose the next job that will fit well in terms of the processor allocations. The results show that the genetic algorithm performs better in terms of the percentage deviation of the solution from the lower bound value when new crossover operator is used along with the insertion mutation. Some of the results from this study are included in this paper for comparison.

5. Other heuristic methods

The ant colony system (Dorigo, 1997) is another popular algorithm which is widely used in optimisation problems. Recently, Ying & Lin (2006) applied ant colony system (ACS) to hybrid flow-shops with multiprocessors tasks. Authors determine the jobs-permutation at the first stage, by ACS approach. Other stages are scheduled using an ordered list which is obtained by referring to completion times of jobs at the previous stage. Authors also apply the same procedure to the inverse problem to obtain the backward schedules. After that they employ a local search approach to improve the best schedule obtained in current iteration. Their computational results show that ACS has better performance compared to TS or GA though their algorithm is not any simpler than that of TS or GA.

Recently, Tseng & Liao (2008) tackled the problem by using particle swarm optimization. Their algorithm differs in terms of encoding scheme to construct a particle, the velocity equation and local search mechanism when compared to the basic PSO and the hybrid PSO algorithms presented here. Based on their published experimental results, PSO algorithm developed by Tseng & Liao (2008) performs well in this scheduling problem. Lately, Ying (2008) applied iterated greedy (IG) heuristic in search of a simpler and more efficient solution. The IG heuristic also shows a notable performance as it's tailored to this particular problem.

6. Experimental results

The performance of all the meta-heuristics described above is tested using intensive computational experiments. Similarly, performance of the basic PSO and the hybrid PSO algorithms, in minimizing the overall completion time of all jobs, is also tested using the same computational experiments. The effects of various parameters such as number of jobs and processor configurations on the performance of the algorithm are also investigated. The results are presented in terms of Average Percentage Deviation (APD) of the solution from the lower bound which is expressed as:

$$APD = \frac{C_{\max} - LB}{LB} \times 100 \tag{3}$$

Here, C_{max} indicates the completion time of the jobs and LB indicates the lower bound calculated for the problem instance. The lower bounds used in this performance study were developed by Oğuz et al. (2004) and it is given with the following formula:

$$LB = \left\{ \max_{i \in M} \left\{ \min_{j \in J} \sum_{l=1}^{i-1} t_{l,j} \right\} + \frac{1}{m_i} \sum_{j \in J} p_{i,j} \times t_{i,j} + \min_{j \in J} \left\{ \sum_{l=i+1}^{l} t_{l,j} \right\} \right\} \tag{4}$$

In the above formula, M and J represent the set of stages and set of jobs consecutively. We used the benchmark data available at Oğuz's personal web-site (http://home.ku.edu.tr/~coguz/). Data set contains instances for two types of processor configurations:

(i) **Random processor:** In this problem set, the number of processors in each stage is randomly selected from a set of {1,..., 5}

(ii) **Fixed processor:** In this case identical number of processors assigned at each stage which is fixed to 5 processors.

For both configurations, a set of 10 problem instances is randomly produced for various number of jobs (n=5, 10, 20, 50, 100) and various number of stages (k=2, 5, 8). For each n and k value, the average APD is taken over 10 problem instances.

Table 2 and 3 presents the APD results obtained for the basic PSO and the hybrid PSO algorithms. Furthermore, we compare the results with genetic algorithm developed by Oğuz and Ercan (2005), tabu search by Oğuz et al. (2004), ant colony system developed by Ying and Lin (2006), iterated greedy algorithm (IG) by Ying (2008) and PSO developed by Tseng & Liao (2008). The performance of GA (Oğuz and Ercan,2005) is closely related to the control parameters and the cross over and mutation techniques used. Therefore, in Tables 2 and 3, we include the best results obtained from four different versions of GA reported. The performance comparison given in below tables is fair enough as most of the authors were employing the same problem set. Furthermore, all the algorithms use the same LB. However there are two exceptions. For the GA, authors use an improved version of the LB than the one given in equation 4. In addition, the PSO developed by Tseng & Liao (2008) is tested with different set of problems and with the same LB as in GA. However, these problems

also have the same characteristic in terms of number of stage, generation methods for processor and processing time requirements, etc.
From the presented results in Table 2 and 3, it can be observed that TS delivers reasonably good results only in two stage case; whereas GA demonstrates a competitive performance for small to medium size problems. For large number of jobs (such as n=50, 100) and large number of stages (k=8), GA did not outperform ACS, IG or PSO. When we compare ACS with TS and GA, we can observe that it outperforms TS and GA in most of the cases. For instance, it outperforms GA in 8 out of 12 problems in random processor case (Table 2). Among those, the performance improvement was more than %50 in six cases. On the other hand, IG gives a better result when compared to ACS in all most all the cases. The IG heuristic shows notable performance improvement for large problems (n=50 and n=100). For example, in n=100 and k=8 case, IG result is %71 better as compared to GA, %95 compared to TS and %7 compared to ACS.

The basic PSO algorithm presented here approximates to GA and ACS results though it did not show a significant performance improvement. PSO outperformed GA 4 in 12 problems for random processors and 1 in 12 problems for fixed processors. The best performance improvement was 54%. On the other hand, PSO-SA hybrid outperformed GA 7 and ACS 3 in 12 problems. In most of the cases, PSO-SA and PSO-TS outperformed the basic PSO algorithm. Amongst the two hybrids experimented here, PSO-SA gave the best results. The best result obtained with PSO-SA was in 50-jobs, 5-stages case, where the improvement was about 59% when compared to GA but this was still not better than ACS or IG. However, PSO developed by Tseng & Liao (2008) gives much more competitive results. Although thier results are for different set of problems, it can be seen that their algorithm performance improves when the problem size increases. Authors compared their algorithm with GA and ACS using the same set of data and reported that their PSO algorithm supersedes them, in particular for large problems. From the results, it can also be observed that when the number of processors are fixed, that is m_j =5, the scheduling problem becomes more difficult to solve and APD results are relatively higher. This is evident in the given results of different metaheuristic algorithms as well as the basic PSO and the hybrid PSO algorithms presented here. In the fixed processor case, PSO-SA, which is the best performing algorithm among the three PSO algorithms, outperformed GA in 3 out of 12 problems and the best improvement achieved was %34. The performance of ACS is better for large problems though IG is dominant in most of the problems. For the fixed problem case, PSO algorithm developed by (Tseng & Liao, 2008) did not show an exceptional performance when compared to GA or ACS for smaller problems though for large problems (that is n=50 and 100) their PSO algorithm outperforms all.
The execution time of the algorithms is another indicator of the performance though it may not be a fair comparison as different processors and compilers used for each reported algorithm in literature. For instance, the basic PSO and the hybrid PSO algorithms presented here are implemented using Java language and run on a PC with 2GHz Intel Pentium processor (with 1024 MB memory). GA (Oğuz & Ercan, 2005) implemented with C++ and run on a PC with 2GHz Pentium 4 processor (with 256 MB memory), IG (Ying, 2008) with visual C#.net and PC with 1.5GHz CPU and ACS (Ying & Lin, 2006) with Visual C++ and PC with 1.5 GHz Pentium 4 CPU. However, for the sake of completeness we execute GA, the basic PSO and the hybrid PSO on the same computing platform using one easy (k=2, n=10) and one difficult problem (k=8, n=100) for the same termination criterion of 10000 iterations for all the algorithms. Results are reported in Table 4, which illustrates the speed performance of PSO. It can be seen

that PSO is approximately 48% to 35% faster than reported GA CPU timings. The fast execution time of PSO is also reported by Tseng and Liao (2008). However, in our case hybrid algorithms were as costly as GA due to computations in SA and TS steps.

k	n	TS (Oğuz et al. 2004)	GA (Oğuz & Ercan, 2005)	ACS (Ying & Lin, 2006)	IG (Ying, 2008)	Basic PSO	PSO-SA	PSO-TS	PSO* (Tseng & Liao, 2008)
	10	3.0	1.60	1.60	1.49	2.7	1.7	2.1	2.8*
2	20	2.88	0.80	1.92	1.87	2.88	1.12	1.92	5.40
	50	2.23	0.69	2.37	2.21	2.38	2.4	2.4	2.30
	100	9.07	0.35	0.91	0.89	1.82	0.82	1.1	1.62
	10	29.42	11.89	9.51	8.73	10.33	9.78	10.4	10.45
5	20	24.40	5.54	3.07	2.97	8.6	3.19	4.53	6.04
	50	10.51	5.11	1.51	1.49	3.31	2.06	2.98	1.44
	100	11.81	3.06	1.05	1.03	2.11	1.05	1.77	2.80
	10	46.53	16.14	16.50	13.91	18.23	16.14	17.5	19.01
8	20	42.47	7.98	6.77	5.83	12.03	6.69	7.03	5.76
	50	21.04	6.03	2.59	2.47	5.98	3.0	4.19	2.91
	100	21.50	4.12	1.33	1.23	8.78	2.11	5.22	1.53

Table 2. APD of the algorithms for 10 random instances. Random processors case (m_j~[1,5])
(*) Different problem set

k	n	TS (Oğuz et al. 2004)	GA (Oğuz & Ercan, 2005)	ACS (Ying & Lin, 2006)	IG (Ying, 2008)	Basic PSO	PSO-SA	PSO-TS	PSO* (Tseng & Liao, 2008)
	10	10.82	6.13	12.62	8.85	13.8	10.11	12.8	12.75*
2	20	7.25	7.10	10.73	6.93	10.75	9.59	10.73	6.05
	50	5.80	3.34	8.17	5.66	10.32	7.02	8.82	5.69
	100	5.19	2.87	5.66	5.04	7.43	3.21	6.43	6.56
	10	45.14	11.32	26.09	23.49	29.6	11.32	22.2	19.58
5	20	35.13	10.78	15.11	12.64	19.4	10.77	16.5	12.33
	50	28.64	14.91	13.11	11.29	14.17	13.24	13.86	12.47
	100	26.49	11.02	12.45	10.53	12.45	12.45	12.45	11.49
	10	77.21	25.98	25.14	22.17	30.81	25.83	25.83	33.92
8	20	62.99	24.13	25.18	22.79	26.74	24.34	25.02	24.98
	50	54.25	21.87	22.23	20.71	27.01	23.07	25.11	19.41
	100	36.05	19.46	13.90	12.85	20.39	14.43	17.9	15.93

Table 3. APD of the algorithms for 10 random instances. Fixed processor case (m_j =5)
(*) Different problem set

k	n	GA	PSO	PSO-SA	PSO-TS
2	10	12.14	6.3	13.5	12.94
8	100	2109.9	1388.67	4029.1	3816.3

Table 4. Average CPU time (in seconds) of GA, TS and PSO. Processors setting is m_j=5

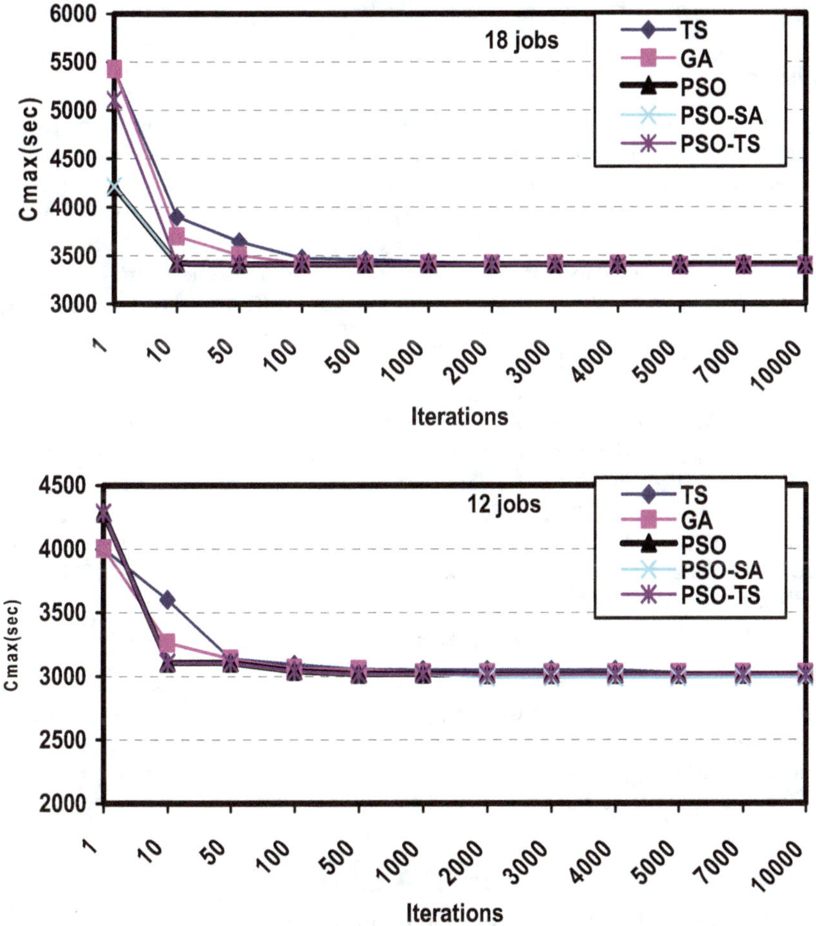

Figure 2. Convergence of TS, GA and PSO algorithms for a machine vision system

Lastly, we run the basic PSO and the hybrid PSO algorithms together with genetic algorithm and tabu search algorithm for the data obtained from a machine-vision system. The system is a multiprocessor architecture designed mainly for machine vision applications. The system comprise of two stages where each stage holds four identical DSP processors from Analog Devices. Data gathered from this system are for 8, 10, 12 and 18 jobs. The number of job is determined by the number of objects to be detected in a given image. The execution time and the processor requirements of parallel algorithms for each job are recorded in order to use them as test problems in our scheduling algorithms. By utilizing this data, which can be obtained from the author, the convergence characteristic of the algorithms is analyzed. As it can be seen from Figure 2, all the algorithms converge very rapidly. However, PSO algorithms converge faster than TS and GA. In addition, PSO-SA finds a slightly better solution for 12 job problem. All the algorithms find a reasonably good solution within 1000 iterations. Hence, for practical application of the scheduling algorithms a small value for termination criteria can be selected.

7. Conclusion

In this chapter, a scheduling problem, defined as hybrid flow-shops with multiprocessor tasks, is presented together with various meta-heuristic algorithms reported for the solution in literature. As the solution to this scheduling problem has merits in practise, endeavour to find a good solution is worthy. The basic PSO and the hybrid PSO algorithms are employed to solve this scheduling problem, as PSO proven to be a simple and effective algorithm applied in various engineering problems. In this particular scheduling problem, a job is made up of interrelated multiprocessor tasks and each multiprocessor task is modelled with its processing requirement and processing time. The objective was to find a schedule in which completion time of all the tasks will be minimal. We observe that basic PSO has a competitive performance as compared to GA and ACS algorithms and superior performance when compared to TS. Considering the simplicity of the basic PSO algorithm, the performance achieved is in fact impressive. When experimented with the hybrids of PSO, it is observed that PSO-SA combination gave the best results. Hybrid methods improved the performance of PSO significantly though this is achieved at the expense of increased complexity. When compared to other published results on this problem, it can be concluded that IG algorithm (Ying, 2008) and PSO given by (Tseng & Liao, 2008) are the best performing algorithms on this problem so far. In terms of effort to develop an algorithm, execution time of algorithm and simplicity to tune it, PSO tops all the other metaheuristics.

As in many practical scheduling problems, it is likely to have precedence constraints among the jobs hence in future study hybrid flow-shops with precedence constraints will be investigated. In addition, PSO may be applied to other scheduling problems and its performance can be exploited in other engineering problems.

8. References

Brucker, P. & Kramer, B. (1995). Shop scheduling problems with multiprocessor tasks on dedicated processors, *Annals of Operations Research*, Vol. 50, 13-27

Caraffa, V.; Ianes, S.; Bagchi, T.P. & Sriskandarajah, C. (2001). Minimizing make-span in blocking flow-shop using genetic algorithms, *International Journal of Production Economics*, Vol. 70, 101-115

Chan, J. & Lee, C. Y. (1999). General multiprocessor task scheduling, *Naval Research Logistics*, Vol. 46, 57-74

Chiang, T. C.; Chang, P.Y. & Huang, Y. M. (2006). Multi-processor tasks with resource and timing constraints using particle swarm optimization, *International Journal of Computer Science and Network Security*, Vol.6, No.4, 71-77

Dauzère-Pérès, S. & Paulli, J. (1997). An integrated approach for modelling and solving the general multiprocessor job-shop scheduling problem using tabu search, *Annals of Operations Research*, Vol. 70, 281-306

Dorigo, M. & Gambardella, L.M.(1997). Ant colony system: a cooperative learning approach to the travelling sales man problem. *IEEE Transaction in Evolutionary Computing*, Vol. 1, 53-66

Drozdowski, M. (1996). Scheduling multiprocessor tasks - an overview, *European Journal of Operational Research*, Vol. 94, 215-230

Ercan, M.F. & Fung, Y.F. (2000). The design and evaluation of a multiprocessor system for computer vision, *Microprocessors and Microsystems*, Vol. 24, 365-377

Ercan, M.F. and Fung, Y.F. (2007). Performance of particle swarm optimisation in scheduling hybrid flow-shops with multi-processor tasks, *Lecture Notes in Computer Science*, Vol. 4706, 309-319

Ercan M. F. (2008). A Performance Comparison of PSO and GA in Scheduling Hybrid Flow-Shops with Multiprocessor Tasks, *ACM Symposium on Applied Computing*, Ceara, Brasil.

Gupta J. N. D. (1988). Two stage hybrid flow shop scheduling problem. *Journal of Operational Research Society*, Vol. 39. No: 4, 359–364.

Holland J. H. (1975). *Adaption in Natural and Artificial Systems*, University of Michigan Press, Ann Arbor

Jdrzęjowicz, J. & Jdrzęjowicz, P. (2003). Population-based approach to multiprocessor task scheduling in multistage hybrid flow shops, *Lecture Notes in Computer Science*, Vol. 2773, 279-286

Jin, S.; Schiavone, G. & Turgut, D. (2008). A performance study of multiprocessor task scheduling algorithms, *Journal of Supercomputing*, Vol. 43, 77–97

Kennedy, J., & Eberhart R. (1995) Particle swarm optimization, *Proceedings of IEEE Int. Conf. on Neural Network*, pp. 1942-1948.

Krawczyk, H. & Kubale, M. (1985). An approximation algorithm for diagnostic test scheduling in multi-computer systems, *IEEE Trans. Computers*, Vol. 34/9, 869-8

Lee, C.Y. & Cai, X. (1999). Scheduling one and two-processors tasks on two parallel processors, *IIE Transactions*, Vol. 31, 445-455

Linn, R. & Zhang, W. (1999). Hybrid flow-shop schedule: a survey, *Computers and Industrial Engineering*, Vol. 37, 57-61 [9]

Liu, B.; Wang, L. & Jin, Y.H. (2005). Hybrid particle swarm optimization for flow shop scheduling with stochastic processing time, *Lecture Notes in Artificial Intelligence*, Vol. 3801, 630-637

Murata, T.; Ishibuchi, H. & Tanaka, H. (1996). Multi-objective genetic algorithm and its application to flow-shop scheduling, *Computers and Industrial Engineering*, Vol. 30, 957-968

Oğuz C. & Ercan M.F. (1997). Scheduling multiprocessor tasks in a two-stage flow-shop environment, *Computers and Industrial Engineering*, Vol. 33, 269-272

Oğuz, C.; Ercan, M.F.; Cheng, T.C.E. & Fung, Y.F. (2003). Heuristic algorithms for multiprocessor task scheduling in a two stage hybrid flow shop, *European Journal of Operations Research*, Vol.149, 390-403

Oğuz, C.; Zinder, Y.; Do., V.; Janiak, A. & Lichtenstein, M. (2004). Hybrid flow-shop scheduling problems with multiprocessor task systems, *European Journal of Operations Research*, Vol.152, 115-131

Oğuz, C. & Ercan, M. F. (2005). A genetic algorithm for hybrid flow-shop scheduling with multiprocessor tasks, *Journal of Scheduling*, Vol. 8, 323-351

Poli, R.; Kennedy, J. & Blackwell, T. (2007). Particle swarm optimization an overview, *Swarm Intelligence*, Vol. 1, No. 3, 33-57.

Salman, A.; Ahmad, I. & Al-Madani, S. (2002). Particle swarm optimization for task assignment problem, *Microprocessors and Microsystems*, Vol. 26, 363-371

Shiau, D.F.; Cheng, S.C. & Huang, Y.M. (2008). Proportionate flexible flow shop scheduling via a hybrid constructive genetic algorithm, *Expert Systems with Applications*, Vol. 34, 1133-1143

Sivanandam, S.N.; Visalakshi, P. and Bhuvaneswari, A. (2007). Multiprocessor scheduling using hybrid particle swarm optimization with dynamically varying inertia, *International Journal of Computer Science & Applications*, Vol. 4, 95-106

Tseng, C.T. & Liao, C.J. (2008), A particle swarm optimization algorithm for hybrid flow-shop scheduling with multiprocessor tasks, *International Journal of Production Research*, Vol. 46, 4655-4670.

Tu, K.; Hao, Z. & Chen, M. (2006). PSO with improved strategy and topology for job shop scheduling, *Lecture Notes in Computer Science*, Vol. 4222, 146-155

Xia, W.J. & Wu, Z.M. (2006). A hybrid particle swarm optimization approach for the job-shop scheduling problem, *International Journal of Advance Manufacturing Technology*, Vol. 29, 360-366

Ying, K.C. & Lin, S.W. (2006). Multiprocessor task scheduling in multistage hybrid flow-shops: an ant colony system approach, *International Journal of Production Research*, Vol. 44, 3161-3177

Ying, K.C. (2008). Iterated greedy heuristic for multiprocessor task scheduling problems, *Journal of the Operations Research Society (online edition)*, 1-8

7

Using Opposition-based Learning with Particle Swarm Optimization and Barebones Differential Evolution

Mahamed G.H. Omran
Gulf University for Science and Technology
Kuwait

1. Introduction

Particle swarm optimization (PSO) (Kennedy & Eberhart, 1995) and differential evolution (DE) (Storn & Price, 1995) are two stochastic, population-based optimization methods, which have been applied successfully to a wide range of problems as summarized in Engelbrecht (2005) and Price et al. (2005).

A number of variations of both PSO and DE have been developed in the past decade to improve the performance of these algorithms (Engelbrecht, 2005; Price et al. 2005). One class of variations includes hybrids between PSO and DE, where the advantages of the two approaches are combined. The barebones DE (BBDE) is a PSO-DE hybrid algorithm proposed by Omran et al. (2007) which combines concepts from the barebones PSO (Kennedy 2003) and the recombination operator of DE. The resulting algorithm eliminates the control parameters of PSO and replaces the static DE control parameters with dynamically changing parameters to produce an almost parameter-free, self-adaptive, optimization algorithm.

Recently, opposition-based learning (OBL) was proposed by Tizhoosh (2005) and was successfully applied to several problems (Rahnamayan et al., 2008). The basic concept of OBL is the consideration of an estimate and its corresponding opposite estimate simultaneously to approximate the current candidate solution. Opposite numbers were used by Rahnamayan et al. (2008) to enhance the performance of Differential Evolution. In addition, Han and He (2007) and Wang et al. (2007) used OBL to improve the performance of PSO. However, in both cases, several parameters were added to the PSO that are difficult to tune. Wang et al. (2007) used OBL during swarm initialization and in each iteration with a user-specified probability. In addition, Cauchy mutation is applied to the best particle to avoid being trapping in local optima. Similarly, Han and He (2007) used OBL in the initialization phase and also during each iteration. However, a constriction factor is used to enhance the convergence speed.

In this chapter, OBL is used to improve the performance of PSO and BBDE without adding any extra parameter. The performance of the proposed methods is investigated when applied to several benchmark functions. The experiments conducted show that OBL improves the performance of both PSO and BBDE.

The remainder of the chapter is organized as follows: PSO is summarized in Section 2. DE is presented in Section 3. Section 4 provides an overview of BBDE. OBL is briefly reviewed in

Section 5. The proposed methods are presented in Section 6. Section 7 presents and discusses the results of the experiments. Finally, Section 8 concludes the chaper.

2. Particle Swarm Optimization

Particle swarm optimization (PSO) is a stochastic, population-based optimization algorithm modeled after the simulation of social behavior of bird flocks. In a PSO system, a swarm of individuals (called *particles*) fly through the search space. Each particle represents a candidate solution to the optimization problem. The position of a particle is influenced by the best position visited by itself (i.e. its own experience) and the position of the best particle in its neighborhood (i.e. the experience of neighboring particles). Particle position, $\mathbf{x}_i$, are adjusted using

$$\mathbf{x}_i(t+1) = \mathbf{x}_i(t) + \mathbf{v}_i(t+1) \tag{1}$$

where the velocity component, v_i, represents the step size. For the basic PSO,

$$v_{i,j}(t+1) = wv_{i,j}(t) + c_1 r_{1,j}(t)(y_{i,j}(t) - x_{i,j}(t)) + c_2 r_{2,j}(t)(\hat{y}_j(t) - x_{i,j}(t)) \tag{2}$$

where w is the inertia weight (Shi & Eberhart, 1998), c_1 and c_2 are the acceleration coefficients, $r_{1,j}$, $r_{2,j} \sim U(0,1)$, $\mathbf{y}_i$ is the personal best position of particle i, and $\hat{\mathbf{y}}_i$ is the neighborhood best position of particle i.

The neighborhood best position $\hat{\mathbf{y}}_i$, of particle i depends on the neighborhood topology used (Kennedy, 1999; Kenedy & Mendes, 2002). If a fully-connected topology is used, then $\hat{\mathbf{y}}_i$ refers to the best position found by the entire swarm. That is,

$$\hat{y}_i(t) \in \{y_0(t), y_1(t), ..., y_s(t)\} = min\{f(y_0(t)), f(y_1(t)), ..., f(y_s(t))\} \tag{3}$$

where s is the swarm size.

The resulting algorithm is referred to as the global best (*gbest*) PSO. A pseudo-code for PSO is shown in Alg. 1.

```
for each particle i ∈ 1,...,s do
  Randomly initialize x_i
  Set v_i to zero
  Set y_i = x_i
endfor
Repeat
  for each particle i ∈ 1,...,s do
    Evaluate the fitness of particle i, f(x_i)
    Update y_i
    Update ŷ using equation (3)
    for each dimension j ∈ 1,...,N_d do
      Apply velocity update using equation (2)
    endloop
    Apply position update using equation (1)
  endloop
Until some convergence criteria is satisfied
```

Algorithm 1. General pseudo-code for PSO

Van den Bergh and Engelbrecht (2006) and Clerc and Kennedy (2002) proved that each particle converges to a weighted average of its personal best and neighborhood best position, that is,

$$\lim_{t \to +\infty} x_{i,j}(t) = \frac{c_1 y_{i,j} + c_2 \hat{y}_{i,j}}{c_1 + c_2}$$

This theoretically derived behavior provides support for the barebones PSO developed by Kennedy (2003). It replaces Eqs. 1 and 2 with

$$x_{i,j}(t+1) = N\left(\frac{y_{i,j}(t) + \hat{y}_j(t)}{2}, \left|y_{i,j}(t) - \hat{y}_j(t)\right|\right)$$

Particle positions are therefore randomly selected from a Gaussian distribution with the mean given as the weighted average of the personal best and global best positions, i.e. the swarm attractor. Note that exploration is facilitated via the deviation, $y_{i,j}(t)-\hat{y}_j(t)$, which approaches zero as t increases. In the limit, all particles will converge on the attractor point. Kennedy (2003) also proposed an alternative version of the barebones PSO where Eqs. 1 and 2 are replaced with

$$x_{i,j}(t+1) = \begin{cases} N\left(\frac{y_{i,j}(t) + \hat{y}_j(t)}{2}, \left|y_{i,j}(t) - \hat{y}_j(t)\right|\right) & \text{if } U(0,1) > 0.5 \\ y_{i,j}(t) & \text{otherwise} \end{cases}$$

Based on the above equation, there is a 50% chance that the *j*-th dimension of the particle dimension changes to the corresponding personal best position. This version of PSO biases towards exploiting personal best positions.

3. Differential Evolution

Differential evolution (DE) is an evolutionary algorithm proposed by Storn and Price (1995). While DE shares similarities with other evolutionary algorithms (EA), it differs significantly in the sense that distance and direction information from the current population is used to guide the search process. DE uses the differences between randomly selected vectors (individuals) as the source of random variations for a third vector (individual), referred to as the *target* vector. Trial solutions are generated by adding weighted difference vectors to the target vector. This process is referred to as the mutation operator where the target vector is mutated. A recombination, or crossover step is then applied to produce an offspring which is only accepted if it improves on the fitness of the parent individual.

The basic DE algorithm is described in more detail below with reference to the three evolution operators: mutation, crossover, and selection.

Mutation: For each parent, $\boldsymbol{x}_i(t)$, of generation t, a trial vector, $\boldsymbol{v}_i(t)$, is created by mutating a target vector. The target vector, $\boldsymbol{x}_{i_3}(t)$, is randomly selected, with $i \neq i_3$. Then, two individuals $\boldsymbol{x}_{i_1}(t)$, and $\boldsymbol{x}_{i_2}(t)$ are randomly selected with $i_1 \neq i_2 \neq i_3 \neq i$, and the difference vector, $\boldsymbol{x}_{i_1}$ - $\boldsymbol{x}_{i_2}$, is calculated. The trial vector is then calculated as

$$\boldsymbol{v}_i(t) = \boldsymbol{x}_{i_3}(t) + F(\boldsymbol{x}_{i_1}(t) - \boldsymbol{x}_{i_2}(t)) \tag{4}$$

where the last term represents the mutation step size. In the above, F is a scale factor used to control the amplification of the differential variation. Note that $F \in (0, \infty)$.

Crossover: DE follows a discrete recombination approach where elements from the parent vector, $\boldsymbol{x}_i(t)$, are combined with elements from the trial vector, $\boldsymbol{v}_i(t)$, to produce the offspring, $\boldsymbol{\mu}_i(t)$. Using the binomial crossover,

$$\mu_{ij}(t) = \begin{cases} v_{ij}(t) & \text{if } U(0,1) < P_r \quad \text{or} \quad j = r \\ x_{ij}(t) & \text{otherwise} \end{cases}$$

where $j = 1, ..., N_d$ refers to a specific dimension, N_d is the number of genes (parameters) of a single chromosome, and $r \sim U(1,..., N_d)$. In the above, p_r is the probability of reproduction (with $p_r \in [0, 1]$).

Thus, each offspring is a stochastic linear combination of three randomly chosen individuals when $U(0, 1) < p_r$; otherwise the offspring inherits directly from the parent. Even when $p_r = 0$, at least one of the parameters of the offspring will differ from the parent (forced by the condition $j = r$).

Selection: DE evolution implements a very simple selection procedure. The generated offspring, $\boldsymbol{\mu}_i(t)$, replaces the parent, $\boldsymbol{x}_i(t)$, only if the fitness of the offspring is better than that of the parent.

4. Barebones Differential Evolution

Both PSO and DE have their strengths and weaknesses. PSO has the advantage that formal proofs exist to show that particles will converge to a single attractor. The barebones PSO utilizes this information by sampling candidate solutions, normally distributed around the formally derived attractor point. Additionally, the barebones PSO has no parameters to be tuned. On the other hand, DE has the advantage of not being biased towards any prior defined distribution for sampling mutational step sizes and its selection operator follows a hill-climbing process. Mutational step sizes are determined as differences between individuals in the current population. One of the problems which both PSO and DE share is that control parameters need to be optimized for each new problem.

The barebones DE combines the strengths of both the barebones PSO and DE to form a new, efficient hybrid optimization algorithm. For the barebones DE, position updates are done as follows:

$$x_{i,j}(t) = \begin{cases} p_{i,j}(t) + r_{2,j} \times (x_{i_1,j}(t) - x_{i_2,j}(t)) & \text{if } U(0,1) > p_r \\ y_{i_3,j}(t) & \text{otherwise} \end{cases} \tag{5}$$

where

$$p_{i,j}(t) = r_{1,j}(t)\, y_{i,j}(t) + (1 - r_{1,j}(t))\, \hat{y}_{i,j}(t) \tag{6}$$

with i_1, i_2, $i_3 \sim U(1,\ldots,s)$, $i_1 \neq i_2 \neq i$, $r_{1,j}$, $r_{2,j} \sim U(0,1)$ and p_r is the probability of recombination.

Referring to Eq. 6, $\boldsymbol{p}_i(t)$ represents the particle attractor as a stochastic weighted average of personal best and global best positions, borrowing from the barebones PSO (Kennedy 2003). Referring to Eq. 5, the mutation operator of DE is used to explore around the current attractor, $\boldsymbol{p}_i(t)$, by adding a difference vector to the attractor. Crossover is done with a randomly selected personal best, $\boldsymbol{y}_{i_3}$, as these personal bests represent a memory of best solutions found by individuals since the start of the search process. Also note that the scale factor is a random variable. Using the position update in Eq. 6, for a proportion of (1- p_r) of the updates, information from a randomly selected personal best, $\boldsymbol{y}_{i_3}$, is used (facilitating exploitation), while for a proportion of p_r of the updates step sizes are mutations of the attractor point, $\boldsymbol{p}_i$ (facilitating exploration). Mutation step sizes are based on the difference vector between randomly selected particles, $\boldsymbol{x}_{i_1}$ and $\boldsymbol{x}_{i_2}$. Using the above, the BBDE achieves a good balance between exploration and exploitation. It should also be noted that the exploitation of personal best positions does not focus on a specific position. The personal best position, $\boldsymbol{y}_{i_3}$, is randomly selected for each application of the position update.

5. Opposition-based Learning

Opposition-based learning (OBL) was first proposed by Tizhoosh (2005) and was successfully applied to several problems (Rahnamayan et al., 2008). Opposite numbers are defined as follows:

Let $x \in [a,b]$, then the opposite number x' is defined as

$$x' = a + b - x$$

The above definition can be extended to higher dimensions as follows:

Let $P(x_1, x_2, \cdots, x_n)$ be an n-dimensional vector, where $x_i \in [a_i, b_i]$ and i=1, 2, …, n. The opposite vector of P is defined by $P'(x_1', x_2', \cdots, x_n')$ where

$$x_i' = a_i + b_i - x_i$$

6. Proposed Methods

In this chapter, OBL is used to enhance the performance of PSO and BBDE without adding any extra parameter. Two variants are proposed as follows:

6.1 Improved PSO (iPSO)

An improved version of PSO is proposed such that in each iteration the particle with the lowest fitness, $\boldsymbol{x}_b$, is replaced by its opposite (the *anti-particle*) as follows,

$$x_{b,j} = LB_j + UB_j - x_{b,j}$$

where $x_{b,j} \in [LB_j, UB_j]$, $j=1,2,\ldots,N_d$ and N_d is the dimension of the problem.
The velocity and personal experience of the anti-particle are reset. The global best solution is also updated. A pseudo-code for *i*PSO is shown in Alg. 2.
The rationale behind this approach is the basic idea of opposition-based learning: if we begin with a random guess, which is very far away from the existing solution, let say in worst case it is in the *opposite location*, then we should look in the *opposite direction*. In our case, the guess that is "very far away from the existing solution" is the particle with the lowest fitness.

The main difference between *i*PSO on one side and the approaches proposed by Han and He (2007) and Wang et al. (2007) on the other side, is that we did not introduce any extra parameter to the original PSO. In addition, *i*PSO uses only OBL to enhance the performance of PSO while (Han & He, 2007; Wang et al. 2007) use OBL combined with other techniques (e.g. Cauchy mutation).

```
for each particle i ∈ 1,...,s do
  for each dimension j ∈ 1,...,N_d do
    x_{i,j} = LB_j + r_j × (UB_j − LB_j)
  endloop
endfor
for each particle i ∈ 1,...,s do
  Set v_i to zero
  Set y_i = x_i
endfor
Repeat
  for each particle i ∈ 1,...,s do
    Evaluate the fitness of particle i, f(x_i)
    Update y_i
    Update ŷ using equation (3)
    for each dimension j ∈ 1,...,N_d do
      Apply velocity update using Eq. (2)
    endloop
      Apply position update using equation (1)
    endloop
  Let x_b be the particle with the lowest fitness
  for each dimension j ∈ 1,...,N_d do
    x_{b,j} = LB_j + UB_j − x_{b,j}
  endloop
  v_b = 0
  y_b = x_b
  if f(x_b) < f(ŷ)
        ŷ = x_b
  endif
Until some convergence criteria is satisfied
```

Algorithm 2. General pseudo-code for *i*PSO

6.2 Improved BBDE (iBBDE)

Similar to *iPSO*, BBDE is modified such that in each iteration the particle with the lowest fitness, x_b, is replaced by its opposite. The personal experience of the anti-particle is also reset. The global best solution is updated.

7. Experimental Results

This section compares the performance of the proposed methods with that of *gbest* PSO and BBDE discussed in Section 2 and 4, respectively. For the PSO algorithms, $w = 0.72$ and $c_1 = c_2 = 1.49$. These values have been shown to provide very good results (van den Berg, 2002). In addition $s = 50$ for all methods. All functions were implemented in 30 dimensions.
The following functions have been used to compare the performance of the different approaches. These benchmark functions provide a balance of unimodal, multimodal, separable and non-separable functions.
For each of these functions, the goal is to find the global minimizer, formally defined as

$$\text{Given } f\text{: } \Re^{N_d} \rightarrow \Re$$

$$\text{find } \boldsymbol{x}^* \in \Re^{N_d} \text{ such that } f(\boldsymbol{x}^*) \leq f(\boldsymbol{x}), \forall \boldsymbol{x} \in \Re^{N_d}$$

The following functions were used:
A. *Sphere* function, defined as

$$f(\boldsymbol{x}) = \sum_{i=1}^{N_d} x_i^2$$

where $\boldsymbol{x}^* = \boldsymbol{0}$ and $f(\boldsymbol{x}^*) = 0$ for $-100 \leq x_i \leq 100$.
B. *Rosenbrock* function, defined as

$$f(x) = \sum_{i=1}^{N_d - 1} \left(100\left(x_i - x_{i-1}^2\right)^2 + \left(x_{i-1} - 1\right)^2\right)$$

where $\boldsymbol{x}^* = (1,1,\ldots,1)$ and $f(\boldsymbol{x}^*) = 0$ for $-2 \leq x_i \leq 2$.
C. *Rotated hyper-ellipsoid* function, defined as

$$f(\boldsymbol{x}) = \sum_{i=1}^{N_d} \left(\sum_{j=1}^{i} x_j \right)^2$$

where $\boldsymbol{x}^* = \boldsymbol{0}$ and $f(\boldsymbol{x}^*) = 0$ for $-100 \leq x_i \leq 100$.
D. *Rastrigin* function, defined as

$$f(\boldsymbol{x}) = \sum_{i=1}^{N_d} \left(x_i^2 - 10\cos(2\pi x_i) + 10 \right)$$

where $\boldsymbol{x}^* = \mathbf{0}$ and $f(\boldsymbol{x}^*) = 0$ for $-5.12 \le x_i \le 5.12$.

E. *Ackley*'s function, defined as

$$f(x) = -20\exp\left(-0.2\sqrt{\frac{1}{30}\sum_{i=1}^{N_d} x_i^2} \right) - \exp\left(\frac{1}{30}\sum_{i=1}^{N_d} \cos(2\pi x_i) \right) + 20 + e$$

where $\boldsymbol{x}^* = \mathbf{0}$ and $f(\boldsymbol{x}^*) = 0$ for $-32 \le x_i \le 32$.

F. *Griewank* function, defined as

$$f(\boldsymbol{x}) = \frac{1}{4000}\sum_{i=1}^{N_d} x_i^2 - \prod_{i=1}^{N_d} \cos\left(\frac{x_i}{\sqrt{i}} \right) + 1$$

where $\boldsymbol{x}^* = \mathbf{0}$ and $f(\boldsymbol{x}^*) = 0$ for $-600 \le x_i \le 600$.

G. *Salomon* function, defined as

$$f(x) = -\cos\left(2\pi \sum_{i=1}^{N_d} x_i^2 \right) + 0.1\sqrt{\sum_{i=1}^{N_d} x_i^2} + 1$$

where $\boldsymbol{x}^* = \mathbf{0}$ and $f(\boldsymbol{x}^*) = 0$ for $-100 \le x_i \le 100$.

Sphere, Rosenbrock and Rotated hyper-ellipsoid are unimodal, while Rastrigin, Ackley, Griewank and Salomon are difficult multimodal functions where the number of local optima increases exponentially with the problem dimension.

The results reported in this section are averages and standard deviations over 30 simulations. In order to have a fair comparison, each simulation was allowed to run for 50,000 evaluations of the objective function.

Table 1 summarizes the results obtained by applying the two PSO approaches to the benchmark functions. In general, the results show that *i*PSO performed better than (or equal to) *gbest* PSO. Figure 1 illustrates results for selected functions. The figure shows that *i*PSO generally reached good solutions faster than PSO. Similarly, Table 2 shows that *i*BBDE generally outperformed BBDE. Figure 2 illustrates results for selected functions. Thus, Tables 1 and 2 suggest that using the simple idea of replacing the worst particle is the main reason for improving the performance of PSO and BBDE. In additon, we can conclude that opposition-based learning improved the performance of both PSO and BBDE without requiring any extra parameter.

	PSO	*iPSO*
Sphere	0(0)	0(0)
Rosenbrock	22.191441 (1.741527)	20.645323 (0.426212)
Rotated hyper-ellipsoid	2.021006 (1.675313)	0.355572 (0.890755)
Rastrigin	48.487584 (14.599249)	27.460845 (11.966896)
Ackley	1.096863 (0.953266)	0(0)
Griewank	0.015806 (0.022757)	0.006163 (0.009966)
Salomon	0.446540 (0.122428)	0.113207 (0.034575)

Table 1. Mean and standard deviation (±SD) of the function optimization results

	BBDE	*iBBDE*
Sphere	0(0)	0(0)
Rosenbrock	25.826400 (0.216660)	25.942146 (0.209437)
Rotated hyper-ellipsoid	15.409460 (20.873456)	0.905987 (1.199178)
Rastrigin	34.761833 (28.598884)	0(0)
Ackley	0(0)	0(0)
Griewank	0.000329 (0.001800)	0(0)
Salomon	0.166540 (0.047946)	0.149917 (0.050826)

Table 2. Mean and standard deviation (±SD) of the function optimization results

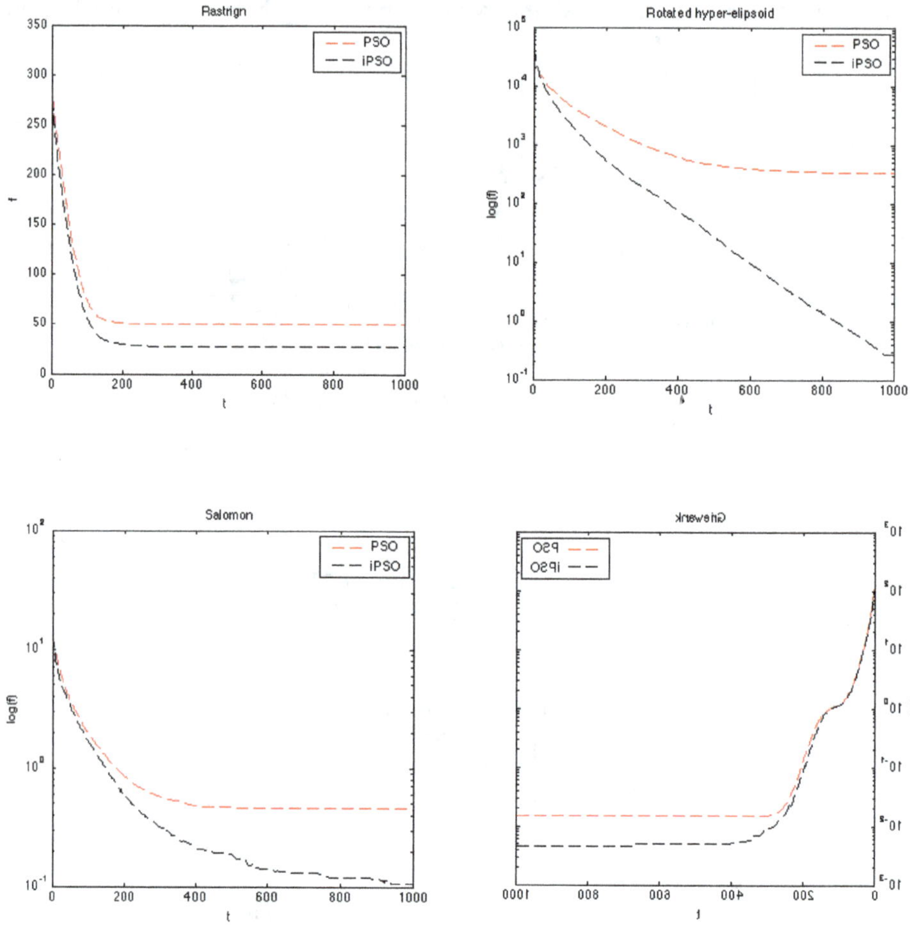

Figure 1. Performance Comparison of PSO and *i*PSO when applied to selected functions

7. Conclusion

Opposition-based learning was used in this chapter to improve the performance of PSO and BBDE. Two opposition-based variants were proposed (namely, *i*PSO and *i*BBDE). The *i*PSO and *i*BBDE algorithms replace the least-fit particle with its anti-particle. The results show that, in general, *i*PSO and *i*BBDE outperformed PSO and BBDE, respectively. In addition, the results show that using OBL enhances the performance of PSO and BBDE without requiring additional parameters. The ideas introduced in this chapter could also be used with any PSO/BBDE variant.

Future research will investigate the effect of noise on the performance of the proposed approaches. Furthermore, a scalability study will be conducted. Finally, applying the proposed approaches to real-world problem will be investigated.

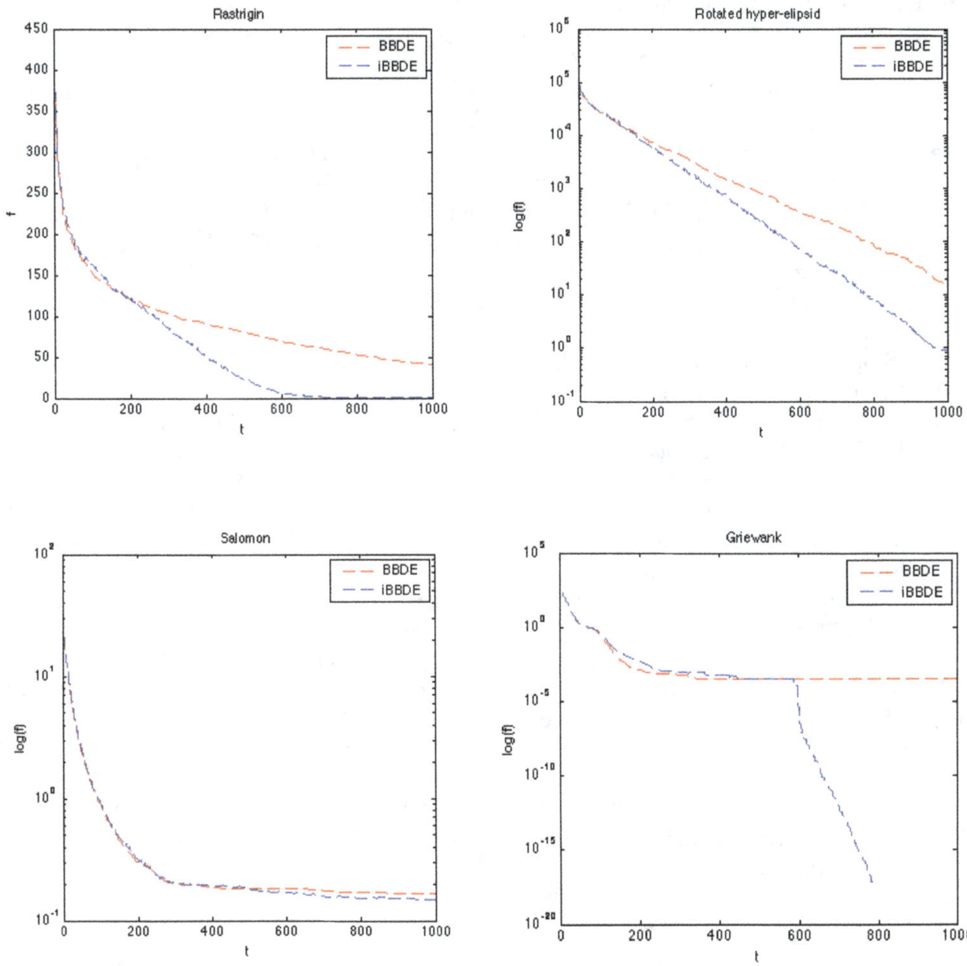

Figure 2. Performance Comparison of BBDE and *i*BBDE when applied to selected functions

8. References

Clerc, M. & Kennedy, J. (2002). The Particle Swarm-Explosion, Stability, and Convergence in a Multidimensional Complex Space. *IEEE Transactions on Evolutionary Computation*, Vol. 6, No. 1, pp. 58-73.

Engelbrecht, A. (2005). *Fundamentals of Computational Swarm Intelligence*, Wiley & Sons.

Han, L. & He, X. (2007). A novel Opposition-based Particle Swarm Optimization for Noisy Problems. *Proceedings of the Third International Conference on Natural Computation*, IEEE Press, Vol. 3, pp. 624 - 629.

Kennedy, J. (1999). Small Worlds and Mega-Minds: Effects of Neighborhood Topology on Particle Swarm Performance. *Proceedings of the IEEE Congress on Evolutionary Computation*, Vol. 3, pp. 1931-1938.

Kennedy, J. (2003). Bare Bones Particle Swarms. *Proceedings of the IEEE Swarm Intelligence Symposium*, pp. 80-87.

Kennedy, J. & Eberhart, R. (1995). Particle Swarm Optimization. *Proceedings of the IEEE International Joint Conference on Neural Networks*, pp. 1942-1948.

Kennedy, J. & Mendes, R. (2002). Population Structure and Particle Performance. *Proceedings of the IEEE Congress on Evolutionary Computation*, pp. 1671-1676, IEEE Press.

Omran, M., Engelbrecht, A. & Salman, A. (2007). Differential evolution based on particle swarm optimization. *Proceedings of the IEEE Swarm Intelligence Symposium*, pp. 112-119.

Price, K.; Storn, R. & Lampinen, J. (2005). *Differential Evolution: A Practical Approach to Global Optimization*, Springer.

Rahnamayan, S.; Tizhoosh, H. & Salama, M. (2008). Opposite-based Differential Evolution. *IEEE Trans. On Evolutionary Computation*, Vol. 12, No. 1, pp. 107-125.

Shi, Y. & Eberhart, R. (1998). A Modified Particle Swarm Optimizer. *Proceedings of the IEEE Congress on Evolutionary Computation*, pp. 69-73.

Storn, R. & Price, K. (1995). Differential evolution - A Simple and efficient adaptive scheme for global optimization over continuous spaces. *Technical Report TR-95-012*, International Computer Science Institute.

Tizhoosh, H. (2005). Opposition-based Learning: A New Scheme for Machine Intelligence. *Proceedings Int. Conf. Comput. Intell. Modeling Control and Autom*, Vol. I, pp. 695-701.

van den Bergh, F. (2002). An Analysis of Particle Swarm Optimizers. *PhD thesis*, Department of Computer Science, University of Pretoria, Pretoria, South Africa, 2002.

van den Bergh, F. & Engelbrecht, A. (2006). A Study of Particle Swarm Optimization Particle Trajectories. *Information Sciences*, Vol. 176, No. 8, pp. 937-971.

Wang, H.; Liu, Y.; Zeng, S.; Li, H. & Li, C. (2007). Opposition-based Particle Swarm Algorithm with Cauchy Mutation. *Proceedings of the IEEE Congress on Evolutionary Computation*, pp. 4750-4756.

8

Personal Best Oriented Particle Swarm Optimizer

Chang-Huang Chen[1], Jonq-Chin Hwang[2] and Sheng-Nian Yeh[2]
[1]*Tungnan University*, [2]*National Taiwan University of Science and Technology*
Taipei, Taiwan, ROC

1. Introduction

Optimization problems are frequently encountered in many engineering, economic or scientific fields that engineers or researchers are seeking to minimize cost or time, or to maximize profit, quality or efficiency, of a specific problem. For example, economic dispatch of power generation, optimal allocation of resources for manufacture, design optimal plant to maximize production, and so many which are unable to enumerate completely. In addition, many optimization problems are very complex and hard to solve by conventional gradient-based techniques, particularly the objective function and constraint are not in closed forms. Thus, the development of a good optimization strategy or algorithm is of great value.

In the past decade, particle swarm optimization (PSO) algorithm [Eberhart & Kennedy 1995, Kennedy and Eberhart 1995] attracts many sights around the world due to its powerful searching ability and simplicity. PSO simulates the swarm behavior of birds flocking and fish schooling that swarms work in a collaborative manner to search for foods as efficient and quick as possible. There are three different types of PSO which are frequently encountered in literature. They are constriction type PSO, constant inertia weight PSO and linearly decreasing inertia weight PSO. Each of them has been successfully applied to many optimization problems.

While empirical studies have proven PSO's usefulness as an optimization algorithm, it does not always fit all problems. Sometimes, it may also get stuck on local optimal. In order to improve the performance, many variants of PSO have been proposed. Some of the proposed algorithms adopted new operations and some of the modifications hybridized with other algorithm. Although they are claimed better than original PSO algorithm, most of them will introduce extra mathematical or logical operations, which, in turn, making algorithm more complicate and spending more computing time. Especially, they, in general, did not present any theoretical models to describe its behavior and support such modifications.

Many researchers have devoted to study how PSO works. They intended to discover the implicit properties of PSO and its weakness and strength via theoretical analysis. The first attempt to analysis PSO is made by Kenndey [Kennedy, 1998]. Meanwhile, Ozcan and Mohan showed that a particle in a simple one-dimensional PSO system follows a path defined by a sinusoidal wave with random amplitude and frequency. However, the effects of inertia weight are not addressed in that paper [Ozcan & Mohan, 1999]. In order to analyze the dynamics of PSO, Yasuda proposed a generalized reduced model of PSO accounting for the inertia weight. The stability analysis is carried out on the basis of both the eigenvalue analysis and numerical simulation [Yasuda et al., 2003]. Trelea has carried out a convergence analysis of PSO and then derived a graphical parameter selection guideline to facilitate choosing parameters [Trelea, 2003].

A formal analysis of the PSO is carried out by Clerc and Kennedy [Clerc & Kennedy, 2002]. By treating the random coefficients as constants, the analysis started from converting the standard stochastic PSO to a deterministic dynamical system. The resulting system was a second-order linear dynamic system whose stability depended on the system's eigenvalues. The parameter space that guarantees stability is also identified. A similar analysis based on deterministic model of the PSO was also carried out in identifying regions in the parameter space that guarantee stability [van den Berg, 2002]. Recently, stochastic convergence analysis of the standard PSO is reported in [Jian, 2007], where a parameter selection guide is also provided to ensure the convergence.

Similar to genetic algorithm or evolutionary algorithm, PSO is also a population-based optimization technique. PSO searches for optimal solution via collaborating with individuals within a swarm of population. Each individual, called particle, is made of two parts, the position and velocity, and proceeds according to two major operations, velocity and position updating rules. Position and velocity represent the candidate solution and step size, a particle will advance in the next iteration, respectively. For an n-dimensional problem and a swarm of m particles, the i-th particle's position and velocity, in general, are denoted as $x_i=[x_{i1}, x_{i2}, \ldots, x_{in}]^T$ and $v_i=[v_{i1}, v_{i2}, \ldots, v_{in}]^T$, for i=1, 2, …, m, respectively, where m is the number of particles, and superscript T stands for the transpose operator. Considering on the inertia weight PSO, the operations of position and velocity are expressed as:

$$v_{id}(t+1) = \omega \cdot v_{id}(t) + c_1 \cdot rnd(1) \cdot (p_{gd} - x_{id}(t)) + c_2 \cdot rnd(1) \cdot (p_{id} - x_{id}(t)) \tag{1}$$

$$x_{id}(t+1) = x_{id}(t) + v_{id}(t+1) \tag{2}$$

where ω is the inertia weight; c_1 and c_2 are two positive constants called acceleration constants; rnd(1) is a uniform random number in (0,1); d is the index of dimension; p_g and p_i are the best position ever found by all particles and the best position a particle ever found so far, respectively; t is the iteration count. Hereafter, p_g and p_i will be called the global best and personal best particle of the swarm, respectively, in this chapter.

Personal best oriented particle swarm optimizer (PPSO), also a varient of particle swarm optimization, is a newly developed optimization solver [Chen & Yeh, 2006a]. PPSO uses the same velocity updating rule as PSO. However, the position updating rule is replaced by (3).

$$x_{id}(t+1) = p_{id} + v_{id}(t+1) \tag{3}$$

The modification came from the observation that since p_{id} is the best particle ever found so far, it may locate around the vinicity of the optimal solution. Fotunately, previous studies showed that PPSO performs well both in testing on a suite of benchmark functions and applying to economic dispatch problems of the power system and others [Chen & Yeh, 2006a, 2006b, 2007, 2008]. However, all the results were obtained from emperical studies. The main drawback of PPSO may lie in a fragile theory basis at first galance. No theoretical

analysis has been found in literature, although it not only can help reveal the behavior of particles but also the convergent property of the proposed approach.
This chapter presents a therotical analysis of PPSO. Based on the analysis, the implicit behavior of PPSO will be revealed. Meanwhile, it also provide a guideline for parameters selection. It is well-known that the particle's trajectory of PSO is characterized by a second-order difference equation. However, it will be clear later that a first order difference equation is sufficient to characterize the particle's behavior of PPSO. A simple mathematical model can help one easily grasp the features of PPSO and convergent tendency.

2. Analysis of PPSO

This section intends to construct the mathematical model of PPSO and, based on the developed model, study the property of PPSO. The analysis is started from a simplified deterministic model, a single particle and one dimension case, keeping all the parameters constants. After that, assuming some parameters be uniformly distributed random numbers, a stochastic model is then built to simulate the nature of PPSO in the next section.

2.1 Simple PPSO: one particle and one dimension case

Each particle in PPSO represents a candidate solution of a specific problem. In other words, a particle is a position in a multidimensional search space in which each particle attempts to explore for an optimal solution with respect to a fitness function or objective function. In this section, a simple PPSO is first derived on which the following analysis is based.
The canonical form of PPSO is represented as .

$$v_{id}(t+1) = \omega \cdot v_{id}(t) + \varphi_1 \cdot (p_{gd} - x_{id}(t)) + \varphi_2 \cdot (p_{id} - x_{id}(t)) \tag{4}$$

$$x_{id}(t+1) = v_{id}(t+1) + p_{id} \tag{5}$$

where $\varphi_1 = c_1 \cdot rnd(1)$ and $\varphi_2 = c_2 \cdot rnd(1)$ are two random numbers drawn uniformly from $(0, c_1)$ and $(0, c_2)$, respectively.
Since (4) and (5) operate on each dimension exclusively, for notation simplicity, one can omit the subscript i and d of x_{id} and v_{id}, and retain subscript g and i of p_g and p_i to emphasize the difference between them. The analysis given below considers only one particle and one dimensional case. However, the results can easily be extended to multidimensional case without losing generality.
Eqations (4) and (5) can now be rewritten as:

$$v(t+1) = \omega \cdot v(t) + \varphi_1 \cdot (p_g - x(t)) + \varphi_2 \cdot (p_i - x(t)) \tag{6}$$

$$x(t+1) = v(t+1) + p_i \tag{7}$$

Substituting (6) into (7), one has:

$$x(t+1) = \omega \cdot v(t) + \varphi_1 \cdot (p_g - x(t)) + \varphi_2 \cdot (p_i - x(t)) + p_i \tag{8}$$

Since

$$x(t) = v(t) + p_i \tag{9}$$

It has

$$v(t) = x(t) - p_i \tag{10}$$

and

$$x(t+1) = \omega \cdot (x(t) - p_i) + \varphi_1 \cdot (p_g - x(t)) + \varphi_2 \cdot (p_i - x(t)) + p_i \tag{11}$$

Rearranging (11), it becomes

$$x(t+1) - (\omega - \varphi)x(t) = \varphi_1 \cdot p_g + (1 + \varphi_2 - \omega) \cdot p_i \tag{12}$$

where $\varphi = \varphi_1 + \varphi_2$. Obviously, a first-order linear difference equation is sufficient to characterize the dynamic behaviors of the simple PPSO.

2.2 Deterministic model of PPSO

Now assume that both p_i and p_g are constants. Also assume that φ_1 and φ_2 are two constants. It turns out that the PPSO becomes a deterministic model described by a first-order linear difference equation with constant coefficients. If the right-hand side of (12) is nonzero, it is a nonhomogeneous linear difference equation. The total solution of a nonhomogeneous linear difference equation with constant coefficients is the sum of two parts, the homogeneous solution, which satisfies the difference equation when the right-hand side of the equation is zero, and the particular solution, which satisfies the difference equation with a nonzero function F(t) on the right-hand side.

The homogeneous solution of a difference equation with constant coefficient is of the form $A\lambda^t$, where λ is called the characteristic root of the difference equation and A is a constant to be determined by the boundary (initial) condition.

The homogeneous solution and particlar solution of (12) can be obtained readily

$$x_h(t) = A(\omega - \varphi)^t \tag{13}$$

and

$$x_p(t) = [\varphi_1 \cdot p_g + (1 + \varphi_2 - \omega) \cdot p_i]/(1 + \varphi - \omega) \tag{14}$$

Here, subscript h and p are used to denote the homogeneous solution and particular solution. The total solution of (12) become

$$x(t) = A(\omega - \varphi)^t + p_w \tag{15}$$

where

$$p_w = [\varphi_1 \cdot p_g + (1 + \varphi_2 - \omega) \cdot p_i]/(1 + \varphi - \omega) \tag{16}$$

is called the weighted mean of p_g and p_i. Given the initial condition $x(0)=x_0$, the dynamic property of a particle is completely characterized by

$$x(t) = (x_0 - p_w)(\omega - \varphi)^t + p_w \tag{17}$$

where x_0 is the initial vale of x(t) and A = (x_0 - p_w). Equation (12) and (17) represent the position or trajectory that a particle may explore in implicit and explicit form.

2.2.1 Convergence property of the deterministic PPSO

Apparently, if (ω - ϕ) satisfies the following condition

$$|(\omega - \varphi)| < 1 \tag{18}$$

or

$$-1 < (\omega - \varphi) < 1 \tag{19}$$

Then

$$\lim_{t \to \infty} (x(t)) = p_w \tag{20}$$

The limit does exist whenever p_w is an arbitrary point in the search space, i.e., p_w is finite. It is obvious that if $0 < \omega < 1$, it leads to $(1 + \varphi - \omega) > 0$ since $\varphi = \varphi_1 + \varphi_2 > 0$, and the weighted mean p_w is finite.

Hereafter, finite p_w and $0 < \omega < 1$ are assumed, unless stated explicitly. The feasible region in which x(t) is strictly converges for $0<\omega<1$ and $-1<\varphi< 2$ is plotted in Fig.1, where the gray area is the feasible region if stability is concerned, and the dark line on the center corresponds to $\omega = \varphi$.

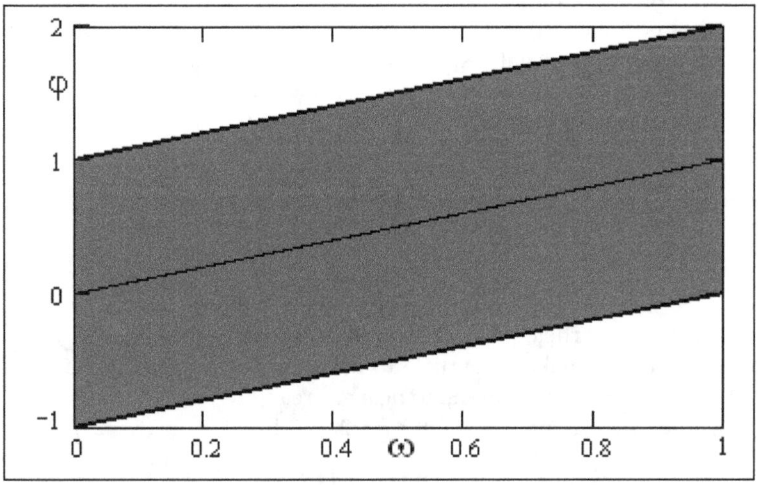

Figure 1. The gray region is the feasible region which particle strictly converges for $0 < \omega < 1$ and $-1< \varphi < 2$. The centered dark-line on the gray area corresponds to $\omega = \varphi$

2.2.2 Step size

The span the particle advances in the next step is calculated using the successive positions at t and (t+1),

$$x(t) = (x_0 - p_w)(\omega - \varphi)^t + p_w \tag{21}$$

and

$$x(t+1) = (x_0 - p_w)(\omega - \varphi)^{t+1} + p_w \tag{22}$$

Define the step size as

$$\begin{aligned} \Delta x(t) &\equiv x(t+1) - x(t) \\ &= (\omega - \varphi - 1)(x_0 - p_w)(\omega - \varphi)^t \end{aligned} \tag{23}$$

Since

$$(x_0 - p_w)(\omega - \varphi)^t = -(p_w - x(t)) \tag{24}$$

It has

$$\Delta x(t) = -(\omega - \varphi - 1) \cdot dx \tag{25}$$

where

$$dx \equiv (p_w - x(t)) \tag{26}$$

is the distance between the current position x(t) and the weighted mean, p_w. Equation (26) tells that the step size is a multiple, defined by $-(\omega - \varphi - 1)$, of the distance between x(t) and p_w. If $-(\omega - \varphi - 1)$ is positive, x(t) moves in aligning with the direction from current position to p_w and, if $-(\omega - \varphi - 1)$ is negative, x(t) moves on the opposite side. The former make particles moving close to p_w and the latter make particles get far way from p_w.
Now, define a step size control factor, δ, as:

$$\delta \equiv -(\omega - \varphi - 1) \tag{27}$$

Then

$$\Delta x(t) = \delta \cdot dx \tag{28}$$

Obviously, how long a particle will advance in next turn is controlled by the step size control factor δ. A large δ makes a particle to move far away from current position and a small value of δ makes a particle moving to nearby area.
Meanwhile, it is interesting to note that if $0 < \delta < 2$, (27) becomes to

$$0 < -(\omega - \varphi - 1) < 2 \tag{29}$$

or

$$-1 < (\omega - \varphi) < 1 \tag{30}$$

This agrees with (19). In other words, if the step size control factor satisfies $0 < \delta < 2$, the deterministic PPSO converges to p_w. Otherwise, it diverges.

Clearly, under condition $0 < \delta < 2$, the deterministic PPSO is stable, otherwise, it is unstable. Since δ is a function of ω and φ, the choices of ω and φ affect the magnitude of the step size control factor, or, in other words, affect the stability of PPSO.
Returning to (30), there are two cases are especially worthy to pay attention:

(a) $0 < (\omega - \varphi) < 1$

This case corresponds to

$$0 < -(\omega - \varphi - 1) < 1 \Rightarrow 0 < \delta < 1 \tag{31}$$

In such situation, x(t+1) moves to the region to the left of p_w whenever p_w is greater than x(t), or the region to the right of p_w whenever p_w is less then x(t).

(b) $-1 < (\omega - \varphi) < 0$

This case corresponds to

$$1 < -(\omega - \varphi - 1) < 2 \Rightarrow 1 < \delta < 2 \tag{32}$$

This means that x(t+1) advances to the region to the right of p_w whenever p_w is less than x(t), or the region to the left of p_w whenever p_w is greater then x(t).
These two cases are illustrated in Figs.2 and 3. It is apparent that the step size control factor affects how far the particle moves. Since the step size is proportional to δ, a large δ corresponds to advancing in a large step and small δ corresponds to small step. Moreover, a positive δ makes x(t) move along the direction from x(t) to p_w ,while a negative δ causes it move along the opposite direction. By controlling δ, or equivalently $(\omega - \varphi)$, particles movement will be totally grasped. It is expected that if δ is uniformly changed in (0, 2), then x(t) will vibrate around the center position, p_w, the weighted midpoint of p_g and p_i. This is a very important property of PPSO. Similar phenomenon has been observed [Kennedy, 2003] and verified theoretically in PSO [Clerc & Kennedy, 2002].

2.2.3 Parameters selection

Equation (27) defines the step size control factor. It provides clues for determining parameters. First, confine the step size control factor within $(\delta_{min}, \delta_{max})$, where δ_{min} and δ_{max} are the lower and upper limits of the step size control factor, respectively, it turns out that

$$\delta_{min} < -(\omega - \varphi - 1) < \delta_{max} \tag{33}$$

After proper rearrangement, (33) becomes

$$(\delta_{min} + \omega - 1) < \varphi < (\delta_{max} + \omega - 1) \tag{34}$$

According to (34), once the lower and upper bounds of the step size control factor are specified, the range of φ depends on ω. The most important is that a stable PPSO requires, based on (29), $\delta_{min} = 0$ and $\delta_{max} = 2$. Substituting these two values into into (34), it has

$$\omega - 1 < \varphi < 1 + \omega \tag{35}$$

Equation (35) says that, if φ uniformly varies from $(\omega - 1)$ to $(\omega + 1)$, x(t) will explore the region from $(p_w - dx)$ to $(p_w + dx)$. It also implies that it is possible to use a negative value of φ while PPSO is still stable. This fact has been shown in Fig.1.

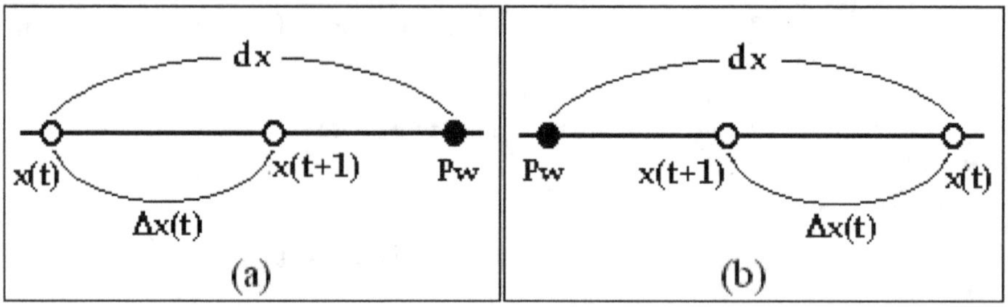

Figure 2. The next position, x(t+1), a particle will move to for $0 < \delta < 1$, (a) $p_w > x(t)$ and (b) $p_w < x(t)$

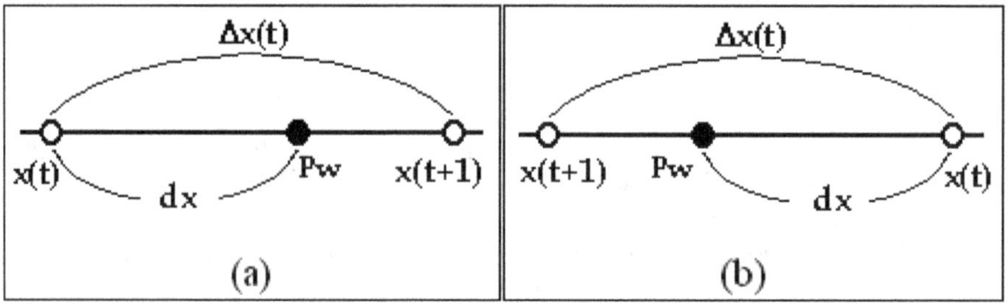

Figure 3. The next position, x(t+1), a particle will move to for $1 < \delta < 2$, (a) $p_w > x(t)$ and (b) $p_w < x(t)$

Since φ_1 and φ_2 are both positive numbers, so is φ. Fig.4 shows the case that φ is positive and is restricted to $0 < \varphi < 2$.
If ω is assigned, from (27), one also has

$$\varphi_{min} + 1 - \omega < \delta < \varphi_{max} + 1 - \omega \tag{36}$$

where φ_{min} and φ_{max} are the lower and upper limits of φ. Thus, one can use (36) to predict the range the particle attempts to explore for a specific range of φ, if ω is given. From (36), one can also readily verify that $\varphi_{min} = \omega - 1$ and $\varphi_{max} = \omega + 1$ result in $\delta = 0$ and 2, respectively, agreeing with (29).
A graph of step size control factor versus ϕ with ω as parameter is plotted in Fig.5 for $0 < \varphi < 2$ and $0 < \omega < 1$. The gray area corresponds to convergent condition since $0 < \delta < 2$. One can use this graph to evaluate whether the selected parameters result in convergence or not.

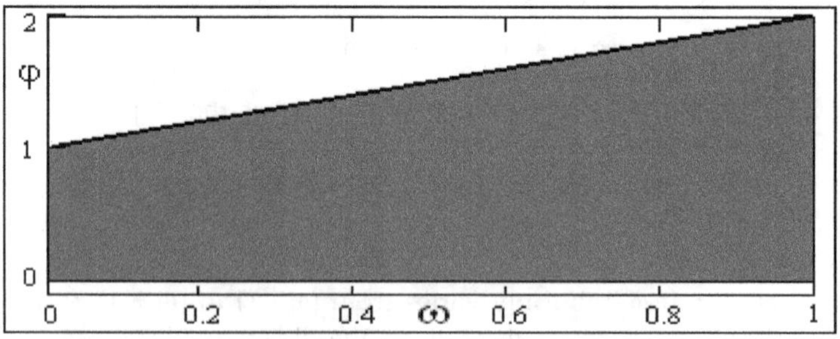

Figure 4. The feasible region for $0 < \omega < 1$ and $0 < \varphi < 2$

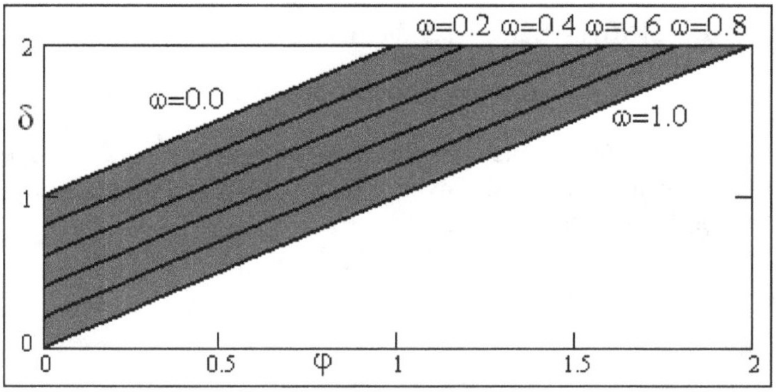

Figure 5. A plot of step size control factor δ versus φ with ω as parameter

2.2.4 Stability

The stability criterion imposed upon the deterministic PPSO is directly obtained from (18), i.e, $|(\omega - \varphi)| < 1$. However, based on (29), an implication of stability means that the step size control factor needs to meet the requirement $0 < \delta < 2$. Hence, stability of the deterministic PPSO can be described by one of the following rules:

(a) $|(\omega - \varphi)| < 1$

or

(b) $0 < -(\omega - \varphi - 1) < 2$

2.2.5 Equilibrium point

According to the above analysis, one can conclude that, for a stable deterministic PPSO, each particle moves in discrete time along the trajectory defined by (12) or (17), with specific step size, and finally settles down at an equilibrium point p_w. The equilibrium point is a function of φ_1 and φ_2. Referring to (16), one can readily verify that if $\varphi_1 > (1 + \varphi_2 - \omega)$, the equilibrium point p_w biases to p_g, and biases to p_i if $\varphi_1 \leq (1 + \varphi_2 - \omega)$. However, the equilibrium point found in PSO is the midpoint of p_g and p_i since $\varphi_1 = \varphi_2$ is usually used in PSO.

3.Stochastic PPSO

Instead of constants, now, restore both φ_1 and φ_2 to be uniform random numbers in $(0, c_1)$ and $(0, c_2)$, respectively. The model of (12) and (17) are still applied except that φ_1 and φ_2 are now two uniform random numbers. Analysis of the dynamic behavior of this stochastic PPSO will be given by extending the analysis provided in the previous section, with the replacement of expectation value for φ_1 and φ_2 as well as x(t) from the probabilistic point of view. In the following analysis, the terms mean value, or simply mean, and expectation value will be used alternatively, in a looser mathematical standard, in the context.

3.1 Convergent property

Considering the explicit representation, Eq.(17), of the trajectory of a particle, since φ_1 and φ_2 are both uniform random numbers, the averaged dynamic behavior of a particle can be observed by its expectation value, i.e.

$$\begin{aligned} E\big(x(t)\big) &= E\big((x_0 - p_w)(\omega - \varphi)^t + p_w\big) \\ &= \big(x_0 - E(p_w)\big)\big(\omega - E(\varphi)\big)^t + E(p_w) \end{aligned} \tag{37}$$

where E(x(t)), $E(p_w)$, and $E(\varphi)$ are the expectation value of x(t), p_w and φ, respectively. Here, φ_1 and φ_2 are assumed to be two exclusive uniform random numbers; and $E(\varphi) = E(\varphi_1) + E(\varphi_2)$. Apparently, if the condition

$$1 < \omega - E(\varphi)) < 1 \tag{38}$$

is true, then

$$\lim_{t\to\infty} E\big(x(t)\big) = E(p_w) \tag{39}$$

According to (39), the trajectory of each particle converges to a random weighted mean, $E(p_w)$, of p_g and p_i, where

$$\begin{aligned} E(p_w) &= E\big([\varphi_1 \cdot p_g + (1+\varphi_2-\omega)\cdot p_i]/(1+\varphi-\omega)\big) \\ &= \frac{E(\varphi_1)\cdot p_g + (1+E(\varphi_2)-\omega)\cdot p_i}{1+E(\varphi_1)+E(\varphi_2)-\omega} \end{aligned} \tag{40}$$

Since

$$1+E(\varphi_1)+E(\varphi_2) > \omega \quad \text{if } 0<\omega<1 \tag{41}$$

$E(p_w)$ is finite for $0 < \omega < 1$, $0 < E(\varphi_1)$ and $0 < E(\varphi_2)$ as well as finite p_i and p_g.
Owing to φ_1 and φ_2 are both uniform random numbers in (0, c_1) and (0, c_2), respectively, it has $E(\varphi_1) = 0.5c_1$, $E(\varphi_2) = 0.5c_2$ and $E(\varphi) = E(\varphi_1 + \varphi_2) = 0.5(c_1 + c_2)$. Thus, (40) becomes

$$\begin{aligned} E(p_w) &= \frac{E(\varphi_1)\cdot p_g + (1+E(\varphi_2)-\omega)\cdot p_i}{1+E(\varphi_1)+E(\varphi_2)-\omega} \\ &= \frac{0.5c_1\cdot p_g + (1+0.5c_2-\omega)\cdot p_i}{1+0.5(c_1+c_2)-\omega} \end{aligned} \tag{42}$$

Obviously, for a stochastic model of PPSO, the random weighted mean p_w is different from that obtained by deterministic model of PSO and PPSO. Meanwhile, for $0 < \omega < 1$, $E(p_w)$ bias to p_i. This means that particle will cluster to p_i instead of p_g.

3.2 Step size

Similar to deterministic PPSO, the step size of the stochastic PPSO can be computed by

$$\begin{aligned} E\big(\Delta x(t)\big) &= E(x(t+1))\text{-}E(x(t)) \\ &= -(\omega - E(\varphi) - 1)(E(p_w) - E(x(t))) \\ &= -(\omega - 0.5(c_1 + c_2) - 1)E(dx) \end{aligned} \tag{43}$$

where

$$E(dx) = E(p_w) - E(x(t)) \tag{44}$$

For a stochastic PPSO, the mean (expectant) step size a particle will move in next turn is computed from (43), which is a multiple of the mean distance between the random weighted mean $E(p_w)$ and mean current position $E(x(t))$. Similar to deterministic PPSO, the mean step size control factor is defined as

$$E(\delta) = -(\omega - E(\varphi) - 1) \tag{45}$$

The step size and step size control factor are no longer static values but stochastic ones. Furthermore, for $0 < E(\delta) < 2$, from (45), it also has

$$-1 < -(\omega - E(\varphi)) < 1 \tag{46}$$

Actually, (46) is the same as (38). Rearranging (46), one has

$$\omega - 1 < E(\varphi) < 1 + \omega \tag{47}$$

Equations (46) and (47) are similar to (30) and (35), respectively, except that the constant φ $(=\varphi_1 + \varphi_2)$ is replaced by sum of the expectation values of two random numbers. As concluded in the previous section, a stable stochastic PPSO equivalently means that the mean step size control factor of each particle's movement must be within the range of $0 < E(\delta) < 2$. In other words, if $E(\varphi)$ lies between $(\omega - 1)$ and $(1 + \omega)$, the system is stable.

3.3 Parameter selection for stochastic PPSO

This subsection discusses how to choose proper parameters for PPSO.

3.3.1 Inertia weight

Recall that $E(\varphi)$ is a positive number since φ is the sum of two uniformly random numbers varying between $(0, c_1)$ and $(0, c_2)$, where c_1 and c_2 are two positive numbers. Now, consider the step size control factor governed by (45) for ω chosen from the following ranges:

(a) $1 < \omega$, it has

$$E(\delta) < E(\varphi) \tag{48}$$

(b) $0 < \omega < 1$, it has

$$E(\varphi) < E(\delta) < 1 + E(\varphi) \tag{49}$$

(c) $-1 < \omega < 0$, it has

$$1 + E(\varphi) < E(\delta) < 2 + E(\varphi) \tag{50}$$

(d) $\omega < -1$, it has

$$2 + E(\varphi) < E(\delta) \tag{51}$$

If $E(\varphi)$ is assigned, Eqs.(48)-(51) specify the average possible value of step size control factor for different choosing ranges of inertia weight ω. For example, if $E(\varphi) = 1.5$, $E(\delta)$ are 1.25, 1.75, 2.75 and 3.75 for ω = 1.25, 0.75, -0.25 and -1.25, respectively. Clearly, it is improper to have a minus value of ω since it will make particle violate the stability rule, i.e., the trajectory of particle diverges.

To have a better vision of parameter selection for $\omega > 1$ and $0 < \omega < 1$, it is better to explain with figure as illustrated in Figs.6 and 7 where the dotted lines represent the domain a

particle may visit in next turn for the cases $\omega = 1.25$ and 0.75 under the condition of $E(\varphi) = 1.5$. The cross sign in the midpoint of two ellipses is the center of the search range. Here, only the case that $E(p_w)$ located to the right of $E(x(t))$ is shown. However, similar aspects can be observed for $E(p_w)$ located to the left of $E(x(t))$.
Since $E(\varphi) = 1.5$, the uniform random number φ varies from 0 to 3. The lower and upper step size control factors are -0.25 and 2.75, respectively, for $\omega = 1.25$. These values are calculated using Eq.(27) . It is seen then in Fig.6 that the search area extends from -0.25E(dx) to 2.75E(dx). Although the upper value of $E(\delta)$ is greater than the upper limit of the step size control factor, the expectation value of the step size control factor is 1.25, which obeys the stability rule given in Eq.(38). From Fig.6, one can also find that if ω is greater than unity, particle is possible to search the region to the left of $E(x(t))$. Meanwhile, the greater ω is, the more the search area shift to left of $E(x(t))$, which will reduce diversity of particle because particles move to the vinicity of $E(x(t))$. Now, refer to Fig.7, for $\omega = 0.75$ and $E(\varphi) = 1.5$,the searach domain are in 0.25E(dx) and 3.25E(dx) with mean of 1.75E(dx). This parameter setting also obeys the stability criterion. It seems both cases of parameter choice is proper. However, refer to Eq.(37), the trajectory of a particle is mainly governed by the term $(\omega - E(\varphi))^t$. If $(\omega - E(\varphi))$ is too small, $E(x(t))$ will vanish quickly and particle may get stuck on local optimum. In other words, the value of $(\omega - E(\varphi))$ represents an index for evaluation of the prematurity of particles. Therefore, it is better to have $0 < \omega < 1$, and empirical studies have shown that it is proper to choice of inertia weight in $0.7 < \omega < 0.8$.

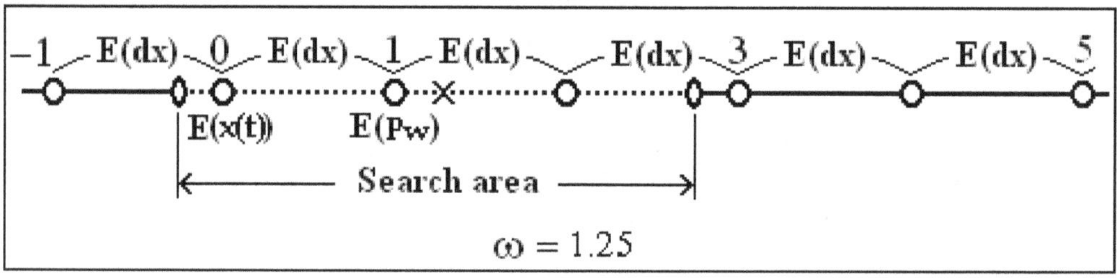

Figure 6. The area the particle will explore for $\omega = 1.25$ and $E(\varphi) = 1.5$

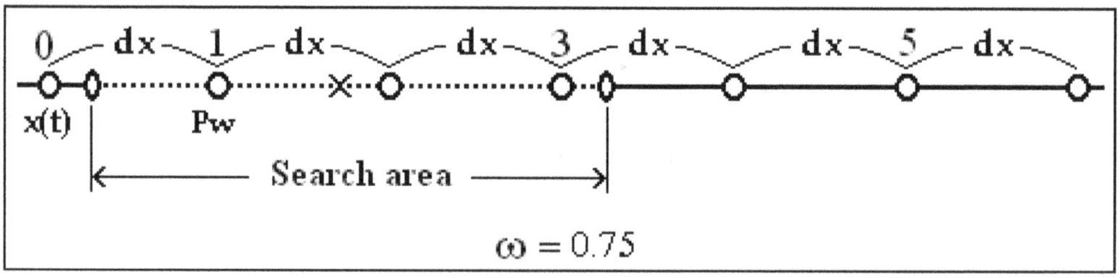

Figure 7. The area the particle will explore for $\omega = 0.75$ and $E(\varphi) = 1.5$

3.3.2 Acceleration coefficient

Recall that c_1 and c_2 are referred to acceleration coefficients, and φ is the sum of two uniform random numbers in $(0, c_1)$ and $(0, c_2)$. The lower and upper limits of φ are then 0 and (c_1+c_2), respectively. To determine c_1 and c_2, it has to consider from three aspects: prematurity, population diversity and particle stability.

If φ varies from 0 to (c_1+c_2) uniformly, from Eq.(27), the lower bound of the step size control factor is determined by the choice of ω, i.e.,

$$\delta_{min} = -\omega + 1 \tag{52}$$

while the upper bound is set by ω, c_1 and c_2, which is

$$\delta_{max} = -\omega + (c_1 + c_2) + 1 \tag{53}$$

For simplicity, it usually has $c_1=c_2=c$, Eq.(53) becomes

$$\delta_{max} = -\omega + 2c + 1 \tag{54}$$

Accounting for stability, in terms of step size control factor, staibility criterion is descibed as

$$0 < E(\delta) < 2 \tag{55}$$

Approximate the expectation value $E(\delta)$ by the average value of δ_{min} and δ_{max}, it has

$$E(\delta) = -\omega + c + 1 \tag{56}$$

Based on (56), one can determine the acceleration coefficients once ω and $E(\delta)$ is assigned. For example, let $\omega = 0.75$ and $E(\delta) = 1.75$ (stisfies Eq.(55)), solve Eq.(56) for c. It is obtained c=1.5. The acceleration coefficients are then set to $c_1=c_2=1.5$. The lower and upper bounds of the step size control factor computed by Eq.(52) and Eq.(54) are 0.25 and 3.25, respectively. The range the particle will search is shown in Fig.7 for this example. It is seen that the search domain stretchs over from 0.25E(dx) to 3.25E(dx), where $E(dx) = E(p_w) - E(x(t))$ is the distance between expectation values of the random weighted mean, p_w, of p_g and p_i and current particle position x(t).

Of course, this is not the unique parameters setting for PPSO. Frequently, it is required to compare the performances between PSO and PPSO. In such situation, the common used parameters for PSO (ω=0.729, $c_1=c_2=1.494$) fit to PPSO since $E(\varphi) = 1.494$, and $E(\delta) = 1.765$ which satisfies Eq.(55).

3.4 Equilibrium point

Both PPSO and PSO define the particles as potential solutions to a problem in a multi-dimensional space with a memory of its ever found best solution and the best solution among all particles. PPSO generates a sequence of x(t) iteratively, and if x(t) is a stable sequence, it has

$$\lim_{t\to\infty} x(t) = E(p_w) \tag{57}$$

where the random weighted mean $E(p_w)$ defined in (42) is the equilibrium point of the sequence. As an optimization solver, it is expected that $E(p_w)$ is the optimum solution. It is seen from (57) that if $p_g = p_i = p$, $E(p_w) = p$. This means that particle settles down at the global best ever found, i.e., PPSO is expected to constantly update the personal best and global best solutions ever found, and finally converges to $E(p_w) = p_g = p_i$, the optimum solution or near optimum solution of the problem at hand.

Note that the random weighted mean of PSO is defined as [Kennedy, 1999 and van den Bergs, 2003]

$$p_w = \frac{c_1 \cdot p_g + c_2 \cdot p_i}{c_1 + c_2} \tag{58}$$

Oviuously, the particles of PSO and PPSO will converge to different equilibrium points. Therefore, in additon to the equilibrium points, the trajectories of PSO and PPSO are also different since trajectories of PPSO and PSO are characterised by a first-order and second-order difference equations [Trelea, 2003, Yasuda et al., 2003, van den Bergh, 2003], respectively. These are the distinctive features of PSO and PPSO.

4. Example Trajectories

To see the properties between PSO and PPSO, the trajectories the particle traversed are investigated by a primitive PSO and PPSO model, where p_g and p_i are set as two arbitrarily constants. To keep thing simple and have a better observation, trajectories of one dimension are considered here. Both the trajectories are generated with same initial condition, i.e., same initial values for position and velocity. Meanwhile, both PSO and PPSO use the same value for the parameters, ω, c_1 and c_2 that they are set as ω=0.729 and c_1=c_2=1.494. Each of the trajectories is constructed by 10000 points and, for fair comparison, each points is generated using the same random numbers for both PSO and PPSO at each time step.

The pseudo-code for generating the trajectories is shown in Fig.8, where x(t) and y(t) are the positions of PSO and PPSO at time step t; vx(t) and vy(t) represent the velocity of PSO and PPSO, respectively; x(0) and y(0) is the initial positions, vx(0) and vy(0) are the initial velocities of PSO and PPSO, respectively.

```
/* pseudo-code for evaluation PSO and PPSO */
Set ω, c1, c2, pg and pi;
Initialize x(0), y(0), vx(0) and vy(0);
For t = 1 to 10000 {
    φ1 = c1 · rnd(1); φ2 = c2 · rnd();
    vx(t) = ω · vx(t − 1) + φ1 · (pg − x(t − 1)) + φ2 · (pi − x(t − 1));
    vy(t) = ω · vy(t − 1) + φ1 · (pg − y(t − 1)) + φ2 · (pi − y(t − 1));
    x(t) = vx(t) + x(t − 1);
    y(t) = vy(t) + pi;
}
/* End */
```

Figure 8. Pseudo-code for evaluating the trajectories of PSO and PPSO

Figure 9 and 10 depicted examples of the trajectories of PSO and PPSO. These plots are obtained with p_g, p_i, x(0), y(0) vx(0) and vy(0) that are arbitrarily set to -50, 10, 0, 0, 20 and 20, respectively. The gray lines in the centre of the figures represent the mean values of x(t) and y(t). They are denoted as μ_x and μ_y for x(t) and y(t), respectively. It is seen obviously that both the trajectories of PSO and PPSO randomly vibrate, or oscillate around the mean values within a limited ranges. The mean values are obtained as μ_x = -20.009, and μ_y = -15.288. These two values very close to the theoretical random weighted mean of p_g and p_i, defined in (58) and (42) for PSO and PPSO, which are calculated to be -20 and -15.394.

Furthermore, the minimum and maximum values of x(t) are -697.131 and 706.212, while the minimum and maximum values of y(t) are -713.624 and 676.268.

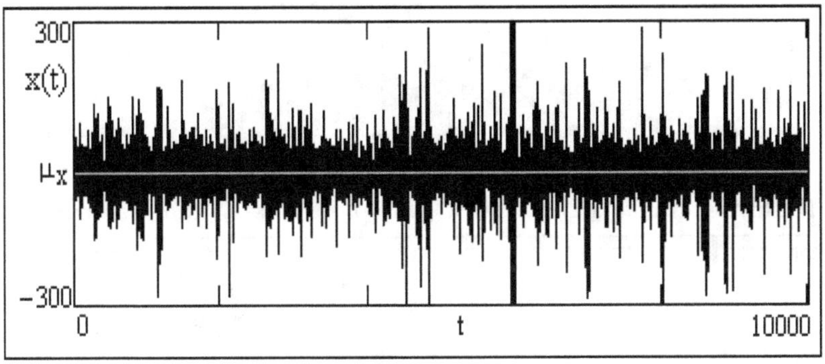

Figure 9. Sample trajectory of x(t) for PSO with p_g=-50, p_i=10, x(0)=0 and vx(0)=20

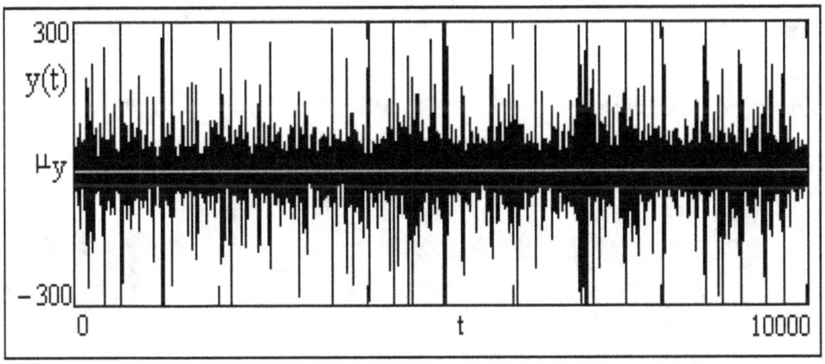

Figure 10. Sample trajectory of y(t) for PPSO with p_g=-50, p_i=10, y(0)=0 and vy(0)=20

Recall that PSO has bell-shaped distribution of the trajectory centred approximately at p_w, i.e., the weighted mean which equals to the midpoint of p_g and p_i [Kennedy, 2003]. This feature also has been observed in PPSO. Refer to Figs.(11) and (12), the histogram plots of the distribution of x(t) and y(t) are illustrated. In these figures, the distributions of the trajectories are drawn in grey lines and the vertical dash-line denoted the mean value of the trajectory. The horizontal and vertical axes represent the values of the trajectory and the occurrences a particle ever explored. The plots of the horizontal axis extend from (μ_x - 3σ_x) to (μ_x + 3σ_x) and (μ_y - 3σ_y) to (μ_y + 3σ_y) for PSO and PPSO, respectively, where σ_x and σ_y are the standard deviations of x(t) and y(t). Obviously, the distribution of the trajectory of the PPSO is also a bell-shaped centred at the random weighted mean. For a comparison, the normal distribution with mean μ_x and standard deviation σ_x for PSO and mean μ_y and standard deviation σ_y for PPSO are drawn in thick solid lines. Clearly, although PSO and PPSO works based on different mathematical models, they have similar dynamic behaviour.

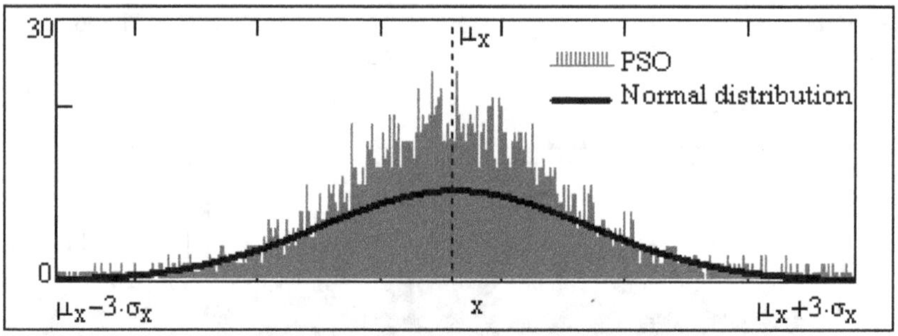

Figure 11. The histogram plot of x(t) for PSO with p_g=-50, p_i=10, x(0)=0 and vx(0)=20

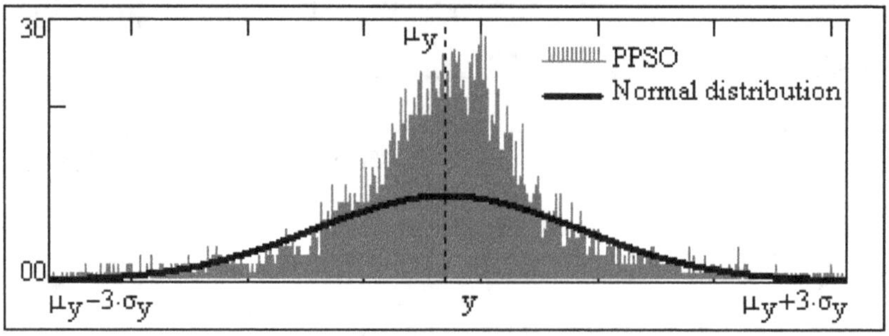

Figure 12. The histogram plot of y(t) for PPSO with p_g=-50, p_i=10, y(0)=0 and vy(0)=20

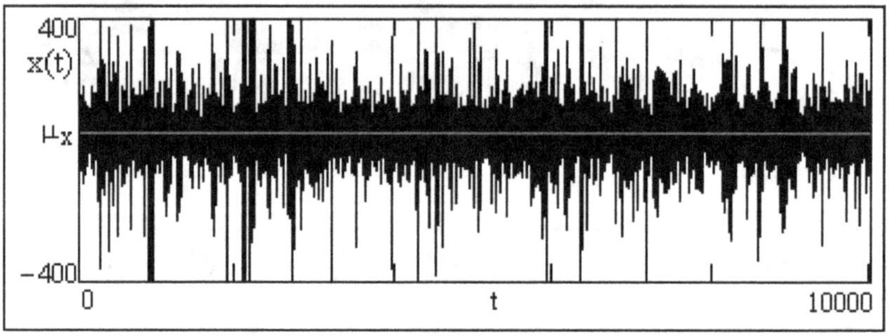

Figure 13. Sample trajectory of x(t) for PSO with p_g = 0, p_i = 100, x(0) = 10 and vx(0) = -2

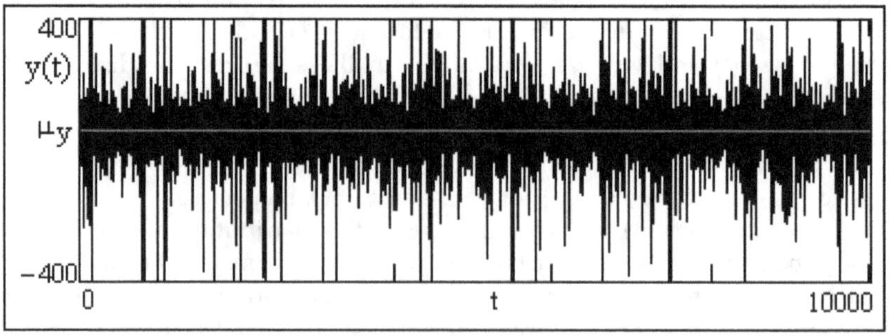

Figure 14. Sample trajectory of y(t) for PPSO with p_g = 0, p_i = 100, y(0) = 10 and vy(0) = -2

Another samples of trajectory for different setting of p_g and p_i as well as initial condition are show in Figs.13 and 14 where p_g, p_i, x(0), y(0) vx(0) and vy(0) are arbitrarily chosen as 0, 100, 10, 10, -2 and -2, respectively. With $p_g = 0$ and $p_i = 100$, the random weighted mean, p_w, of PSO and PPSO are 50 and 57.677. Meanwhile, the mean values, μ_x and μ_y, are 50.588 and 57.609 for PSO and PPSO. The minimum and maximum values are -3.249×10^3 and 3.550×10^3 for x(t) and -1.639×10^3 and 2.941×10^3 for y(t). Apparently, both the trajectories also oscillate around the random weighted mean within a specific domain, which are verified further in the histogram plots shown in Figs.(15) and (16).

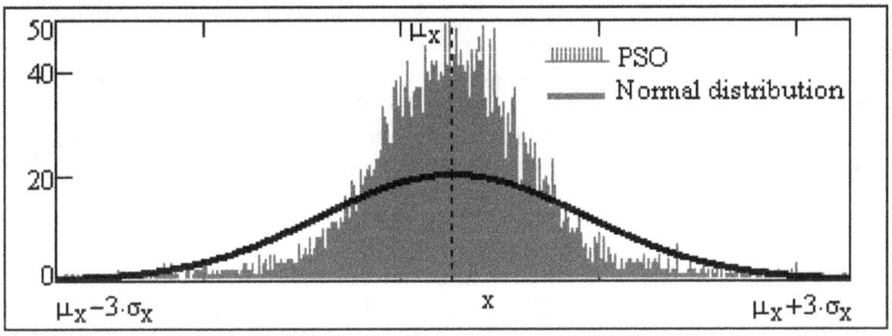

Figure 15. The histogram plot of x(t) for PSO with $p_g = 0$, $p_i = 100$, x(0) = 10 and vx(0) = -2

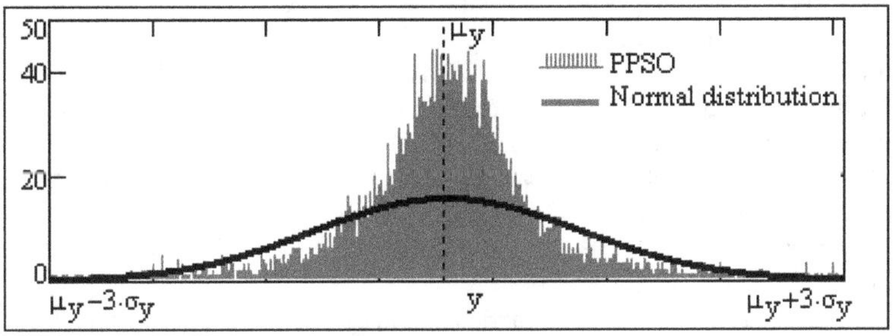

Figure 16. The histogram plot of y(t) for PPSO with $p_g = 0$, $p_i = 100$, y(0) = 10 and vy(0)= -2

5. Conclusion

This chapter intends to provide a theoretical analysis of PPSO to clarify the characteristics of PPSO. The analysis is started from a simplified deterministic model, a single particle and one dimension case, keeping all the parameters constants. After that, assuming the acceleration coefficients as uniformly distributed random numbers, a stochastic model is then built to describe the nature of the PPSO. With the assumption, it is shown that a first-order difference equation is sufficient to describe the dynamic behaviour of the particles. Based on the models, the convergence property is studied and the guidance for parameters selection is provided.

Trajectories comparison between PSO and PPSO are also presented. It is found that, similar to PSO, the particles of PPSO also stochastically explore for optimal solution within a region centered approximately equals to a random weighted mean of the best positions found by an individual (personal best) and its neighbours (global best). Like PSO, bell-shaped distribution of the particle's trajectory is also observed in PPSO. However, the centres of the

distribution of PSO and PPSO are different so that leading to different equilibrium points and, hence, different results and performances.
The results derived in this chapter justify the possibility of PPSO to be an optimization algorithm. Simulation results have been shown that PPSO performs in general better than PSO on a suite of benchmark functions. However, it does not imply that PPSO is a local or global search algorithm even the condition of stability is met. Further research is thus required to improve the search capability.

6. References

Chen, C. H. & Yeh, S. N. (2006a), Personal best oriented constriction type of particle swarm optimization, The *2nd international conference on cybernetics and intelligent systems*, pp. 167-170, Bangkok, Thailand, 7-9 Jun., 2006.

Chen, C. H. & Yeh, S. N. (2006b), Particle swarm optimization for economic power dispatch with valve-point effects, *2006 IEEE PES transmission and distribution conference and exposition : Latin America*, 6 pages, Caracas, Venezuela, 16-19 Aug., 2006.

Chen, C. H. & Yeh, S. N., (2007), Simplified personal best oriented particle swarm optimizer and its applications, The *2nd IEEE Conference on Industrial Electronics and Applications*, pp.2362-2365, Harbin, China, 23-25 May, 2007.

Chen, C. H. & Yeh, S. N. (2008), Personal best oriented particle swarm optimizer for economic dispatch problem with piecewise quadratic cost functions, *International Journal of Electrical Engineering*, vol. 15, no.5, 2008, pp.389-397.

Clerc M. & Kennedy J., (2002), The particle swarm - explosion, stability and convergence in a multidimensional complex space, *IEEE Transaction on Evolutionary Computation*, vol.6, no.1, 2002, pp.58-73.

Eberhart R. C. & Kennedy J. (1995), A new optimizer using particle swarm theory, *Proceedings of the IEEE 6th International Symposium on Micro Machine and Human Science*, pp. 39-43, vol.1, Oct., 1995.

Kennedy J., & Eberhart R. C. (1995), Particle swarm optimization, *Proceedings of the IEEE International Conference on Neural Networks*, pp.1942-1948, vol.4, Perth, Australia, Nov., 1995.

Kennedy J. (1998), The behaviour of particle, *Proceedings of the 7th Annual Conference on Evolutionary Programming*, pp.581-589, 1998.

Kennedy J. (2003), Bare bones particle swarms, *Proceedings of the IEEE Swarm Symposium*, 2003, pp.80-87.

Jiang, M.; Luo Y.P. & Yang S. Y., (2007), Stochastic convergence analysis and parameter selection of the standard particle swarm optimization algorithm, *Information Processing Letters*, vol.102, 2007, pp.8-16.

Ozcan E. & Mohan C. K., (1999), Particle swarm optimization: surfing the waves, *Proceedings of the IEEE Congress on Evolutionary Computation*, pp.1939-1944, vol.3, Jul., 1999.

Trelea I. C., (2003), The particle swarm optimization algorithm: convergence analysis and parameter selection, *Information Processing Letters*, vol.85, 2003, pp.317-325.

van den Bergh F. (2002), Analysis of particle swarm optimization, *Ph.D. dissertation*, University of Pretoria, Pretoria, South Africa, 2002.

Yasuda K., Ide A. & Iwasaki N., (2003), Adaptive particle swarm optimization, pp.1554-1559, *IEEE International Conference on Systems, Man and Cybernetics*, vol.2, Oct., 2003.

9

Particle Swarm Optimization for Power Dispatch with Pumped Hydro

Po-Hung Chen
Department of Electrical Engineering, St. John's University
Taiwan

1. Introduction

Recently, a new evolutionary computation technique, known as particle swarm optimization (PSO), has become a candidate for many optimization applications due to its high-performance and flexibility. The PSO technique was developed based on the social behavior of flocking birds and schooling fish when searching for food (Kennedy & Eberhart, 1995). The PSO technique simulates the behavior of individuals in a group to maximize the species survival. Each particle "flies" in a direction that is based on its experience and that of the whole group. Individual particles move stochastically toward the position affected by the present velocity, previous best performance, and the best previous performance of the group. The PSO approach is simple in concept and easily implemented with few coding lines, meaning that many can take advantage of it. Compared with other evolutionary algorithms, the main advantages of PSO are its robustness in controlling parameters and its high computational efficiency (Kennedy & Eberhart, 2001). The PSO technique has been successfully applied in areas such as distribution state estimation (Naka et al., 2003), reactive power dispatch (Zhao et al., 2005), and electromagnetic devices design (Ho et al., 2006). In the previous effort, a PSO approach was developed to solve the capacitor allocation and dispatching problem (Kuo et al., 2005).

This chapter introduces a PSO approach for solving the power dispatch with pumped hydro (PDWPH) problem. The PDWPH has been reckoned as a difficult task within the operation planning of a power system. It aims to minimize total fuel costs for a power system while satisfying hydro and thermal constraints (Wood & Wollenberg, 1996). The optimal solution to a PDWPH problem can be obtained via exhaustive enumeration of all pumped hydro and thermal unit combinations at each time period. However, due to the computational burden, the exhaustive enumeration approach is infeasible in real applications. Conventional methods (El-Hawary & Ravindranath, 1992; Jeng et al., 1996; Allan & Roman, 1991; Al-Agtash, 2001) for solving such a non-linear, mix-integer, combinatorial optimization problem are generally based on decomposition methods that involve a hydro and a thermal sub-problem. These two sub-problems are usually coordinated by LaGrange multipliers. The optimal generation schedules for pumped hydro and thermal units are then sequentially obtained via repetitive hydro-thermal iterations. A well-recognized difficulty is that solutions to these two sub-problems can oscillate between maximum and minimum generations with slight changes of multipliers (Guan et al., 1994; Chen, 1989). Consequently, solution cost frequently gets stuck at a local optimum rather than at the global optimum. However, obtaining an optimal solution is of priority concern to an electric utility. Even small percentage reduction in production costs typically leads to considerable savings. Obviously, a comprehensive and efficient algorithm for solving the PDWPH problem is still in demand. In the previous efforts, a dynamic programming (DP) approach (Chen, 1989) and a genetic algorithm (GA) technique (Chen & Chang, 1999) have been adopted to solve the PDWPH problem. Although the GA has been successfully applied to solve the PDWPH problem, recent studies have identified some deficiencies in GA performance. This decreased efficiency is apparent in applications in which parameters being optimized are highly correlated (Eberhart & Shi, 1998; Boeringer & Werner, 2004). Moreover, premature convergence of the GA reduces its performance and search capability (Angeline, 1998; Juang, 2004).

This work presents new solution algorithms based on a PSO technique for solving the PDWPH problem. The proposed approach combines a binary version of the PSO technique with a mutation operation. Kennedy and Eberhart first introduced the concept of binary PSO and demonstrated that a binary PSO was successfully applied to solve a discrete binary problem (Kennedy & Eberhart, 1997). In this work, since all Taipower's pumped hydro units are designed for constant power pumping, novel binary encoding/decoding techniques are judiciously devised to model the discrete characteristic in pumping mode as well as the continuous characteristic in generating mode. Moreover, since the basic PSO approach converges fast during the initial period and slows down in the subsequent period and sometimes lands in a local optimum, this work employs a mutation operation that speeds up convergence and escapes local optimums. Representative test results based on the actual Taipower system are presented and analyzed, illustrating the capability of the proposed PSO approach in practical applications.

2. Modeling and Formulation

2.1 List of symbols

DR_i	Down ramp rate limits of thermal unit *i*.
$F_i^t(P_{si}^t)$	Production costs for P_{si}^t.
I_j^t	Natural inflow into the upper reservoir of pumped hydro plant *j* in hour *t*.
N_h	Number of pumped hydro units.
N_s	Number of thermal units.
P_{hj}^t	Power generation (positive) or pumping (negative) of pumped hydro plant *j* in hour *t*.
P_L^t	System load demand in hour *t*.
P_{si}^t	Power generation of thermal unit *i* in hour *t*.
Q_j^t	Water discharge of pumped hydro plant *j* in hour *t*.
$Q_{j,p}^t$	Water pumping of pumped hydro plant *j* in hour *t*.

$R_{hj}^{t}(P_{hj}^{t})$	Spinning reserve contribution of pumped hydro plant *j* for P_{hj}^{t}.
R_{req}^{t}	System's spinning reserve requirements in hour *t*.
$R_{si}^{t}(P_{si}^{t})$	Spinning reserve contribution of thermal unit *i* for P_{si}^{t}.
S_{j}^{t}	Water spillage of pumped hydro plant *j* in hour *t*.
T	Number of scheduling hours.
UR_i	Up ramp rate limits of thermal unit *i*.
V_{j}^{t}	Water volume of the upper reservoirs of plant *j* at the end of hour *t*.
$V_{j,l}^{t}$	Water volume of the lower reservoirs of plant *j* at the end of hour *t*.
v_{i}^{k}	Velocity of particle *i* at iteration *k*.
x_{i}^{k}	Position (coordinate) of particle *i* at iteration *k*.

2.2 Modeling a pumped hydro plant

A pumped hydro plant, which consists of an upper and a lower reservoir, is designed to save fuel costs by generating during peak load hours with water in the upper reservoir, which is pumped up from the lower reservoir to the upper reservoir during light load hours (Fig. 1).

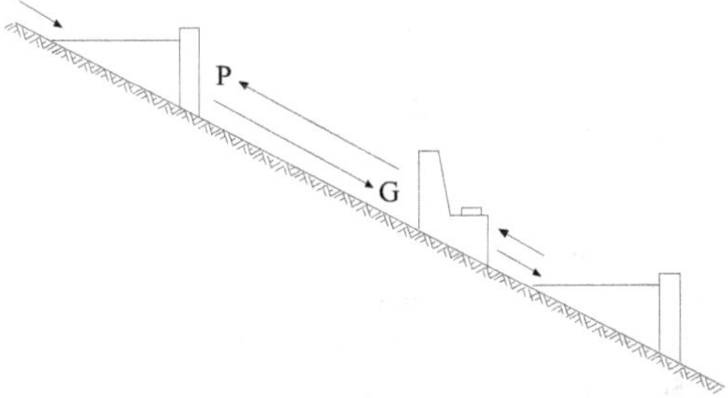

Figure 1. Pumped hydro plant

In generating mode, the equivalent-plant model can be derived using an off-line mathematical procedure that maximizes total plant generation output under different water discharge rates (Wood & Wollenberg, 1996). The generation output of an equivalent pumped hydro plant is a function of water discharged through turbines and the content (or the net head) of the upper reservoir. The general form is expressed as

$$P_{hj}^{t} = f(Q_{j}^{t}, V_{j}^{t-1}) \tag{1}$$

The quadratic discharge-generation function, considering the net head effect, utilized in this work as a good approximation of pumped hydro plant generation characteristics is given as

$$P_{hj}^{t} = \alpha_{j}^{t-1} Q_{j}^{t^2} + \beta_{j}^{t-1} Q_{j}^{t} + \gamma_{j}^{t-1} \tag{2}$$

where coefficients α_j^{t-1}, β_j^{t-1}, and γ_j^{t-1} depend on the content of the upper reservoir at the end of hour *t-1*. In this work, the read-in data includes five groups of α, β, γ coefficients that are associated with different storage volume, from minimum to maximum, for the upper reservoir (first quadrant in Fig. 2). Then, the corresponding coefficients for any reservoir volume are calculated using a linear interpolation (Chen, 1989) between the two closest volume.

In pumping mode, since all Taipower's pumped hydro units are designed for constant power pumping, the characteristic function of a pumped hydro plant is a discrete distribution (third quadrant in Fig. 2).

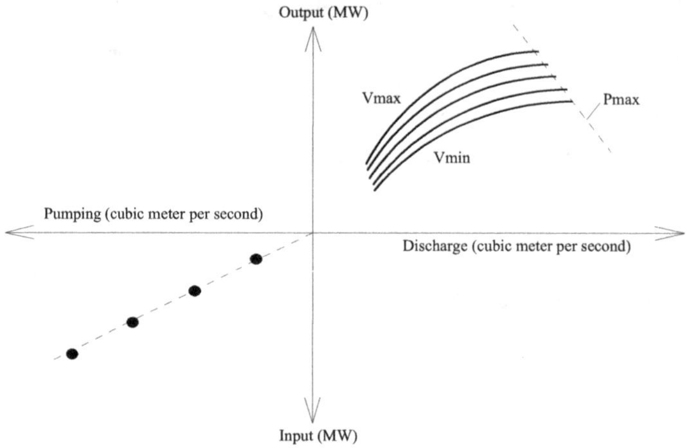

Figure 2. Typical input-output characteristic for a pumped hydro plant

2.3 Objective function and constraints

The pumped hydro scheduling attempts seeking the optimal generation schedules for both pumped hydro and thermal units while satisfying various hydro and thermal constraints. With division of the total scheduling time into a set of short time intervals, say, one hour as one time interval, the pumped hydro scheduling can be mathematically formulated as a constrained nonlinear optimization problem as follows:

$$\textbf{\textit{Problem:}}\ \ \textit{Minimize} \sum_{t=1}^{T}\sum_{i=1}^{N_s} F_i^t(P_{si}^t) \tag{3}$$

Subject to the following constraints:

System power balance

$$\sum_{i=1}^{N_s} P_{si}^t + \sum_{j=1}^{N_h} P_{hj}^t - P_L^t = 0 \tag{4}$$

Water dynamic balance

$$V_j^t = V_j^{t-1} + I_j^t - Q_j^t + Q_{j,p}^t - S_j^t \tag{5}$$

$$V_{j,l}^{t} = V_{j,l}^{t-1} + Q_{j}^{t} - Q_{j,p}^{t} + S_{j}^{t} \tag{6}$$

Thermal generation and ramp rate limits

$$Max(\underline{P_{si}}, P_{si}^{t-1} - DR_i) \le P_{si}^{t} \le Min(\overline{P_{si}}, P_{si}^{t-1} + UR_i) \tag{7}$$

Water discharge limits

$$\underline{Q_j} \le Q_j^t \le \overline{Q_j} \tag{8}$$

Water pumping limits

$$\underline{Q_{j,p}} \le Q_{j,p}^t \le \overline{Q_{j,p}} \tag{9}$$

Reservoir limits

$$\underline{V_j} \le V_j^t \le \overline{V_j} \tag{10}$$

$$\underline{V_{j,l}} \le V_{j,l}^t \le \overline{V_{j,l}} \tag{11}$$

System's spinning reserve requirements

$$\sum_{i=1}^{N_s} R_{si}^t(P_{si}^t) + \sum_{j=1}^{N_h} R_{hj}^t(P_{hj}^t) \ge R_{req}^t \tag{12}$$

3. Refined PSO Solution Methodology

3.1 Basic PSO technique

Consider an optimization problem of D variables. A swarm of N particles is initialized in which each particle is assigned a random position in the D-dimensional hyperspace such that each particle's position corresponds to a candidate solution for the optimization problem. Let x denote a particle's position (coordinate) and v denote the particle's flight velocity over a solution space. Each individual x in the swarm is scored using a scoring function that obtains a score (fitness value) representing how good it solves the problem. The best previous position of a particle is *Pbest*. The index of the best particle among all particles in the swarm is *Gbest*. Each particle records its own personal best position (*Pbest*), and knows the best positions found by all particles in the swarm (*Gbest*). Then, all particles that fly over the D-dimensional solution space are subject to updated rules for new positions, until the global optimal position is found. Velocity and position of a particle are updated by the following stochastic and deterministic update rules:

$$\begin{aligned} v_i^{k+1} = wv_i^k &+ c_1 Rand(\ ) \times (Pbest_i^k - x_i^k) \\ &+ c_2 Rand(\ ) \times (Gbest^k - x_i^k) \end{aligned} \tag{13}$$

$$x_i^{k+1} = x_i^k + v_i^{k+1} \tag{14}$$

where w is an inertia weight, c_1 and c_2 are acceleration constants, and *Rand( )* is a random number between 0 and 1.

Equation (13) indicates that the velocity of a particle is modified according to three components. The first component is its previous velocity, v_i^k, scaled by an inertia, w. This component is often known as "habitual behavior." The second component is a linear attraction toward its previous best position, $Pbest_i^k$, scaled by the product of an acceleration constant, c_1, and a random number. Note that a different random number is assigned for each dimension. This component is often known as "memory" or "self-knowledge." The third component is a linear attraction toward the global best position, $Gbest^k$, scaled by the product of an acceleration constant, c_2, and a random number. This component is often known as "team work" or "social knowledge." Fig. 3 illustrates a search mechanism of a PSO technique using the velocity update rule (13) and the position update rule (14).

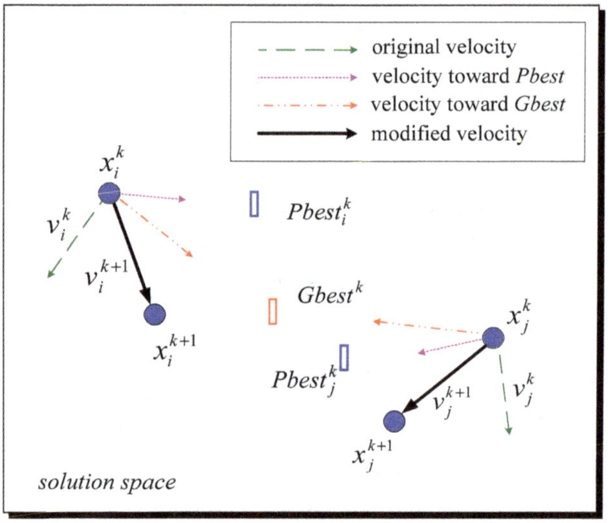

Figure 3. Searching mechanism of a PSO

Acceleration constants c_1 and c_2 represent the weights of the stochastic acceleration terms that push a particle toward *Pbest* and *Gbest*, respectively. Small values allow a particle to roam far from target regions. Conversely, large values result in the abrupt movement of particles toward target regions. In this work, constants c_1 and c_2 are both set at 2.0, following the typical practice in (Eberhart & Shi, 2001). Suitable correction of inertia w in (13) provides a balance between global and local explorations, thereby reducing the number of iterations when finding a sufficiently optimal solution. An inertia correction function called "inertia weight approach (IWA)" (Kennedy & Eberhart, 2001) is utilized in this work. During the IWA, the inertia weight w is modified according to the following equation:

$$w = w_{max} - \frac{w_{max} - w_{min}}{Itr_{max}} \times Itr \tag{15}$$

where w_{max} and w_{min} are the initial and final inertia weights, Itr_{max} is the maximum number of iteration, and Itr is the current number of iteration.

3.2 Binary encoding

For exposition ease, consider a pumped hydro plant with four units. Fig. 4 presents a particle string that translates the encoded parameter-water discharges of each plant into their binary representations.

Hour	1	2	.....	24
	1 0 0 0 1	0 0 0 1 1	.....	1 0 0 1 0

Figure 4. Particle string for a pumped hydro plant with four units

Using a plant's water discharge instead of the plant's generation output, the encoded parameter is more beneficial when dealing with difficult water balance constraints. Each particle string contains 24 sub-strings that represent the solution for hourly discharge/pumping schedules of the pumped hydro plant during a 24-hour period. Each sub-string is assigned the same number of five bits. The first bit is used to identify whether the plant is in generating or pumping mode. The remaining four bits are used to represent a normalized water discharge, q_j^t, in generating mode, or the number of pumping units in pumping mode. In generating mode, the resolution equals $1/2^4$ of the discharge difference from minimum to maximum.

3.3 Decoding of particle string

A particle within a binary PSO approach is evaluated through decoding the encoded particle string and computing the string's scoring function using the decoded parameter. The following steps summarize the detailed decoding procedure.

Step 1. Decode the first bit to determine whether the plant is in generating or pumping mode:

Hour t

b_1	b_2	b_3	b_4	b_5

b_2=b_3=b_4=b_5="0" => "idle mode"
b_1="0" => "pumping mode"
b_1="1" => "generating mode"

Step 2. If in idle mode, P_{hj}^t =0, go to Step 10; if in pumping mode, go to Step 3; if in generating mode, go to Step 6.

Step 3. Decode the remaining four bits of the sub-string to calculate the number of pumping units, N_p^t, and the total volume of pumped water, $Q_{j,p}^t$:

Hour t

0	b_2	b_3	b_4	b_5

$$N_p^t = \sum_{i=2}^{5}(b_i) \qquad b_i \in \{0, 1\} \tag{16}$$

$$Q_{j,p}^t = Q_{j,sp} \times N_p^t \tag{17}$$

where $Q_{j,sp}$ is the constant volume for pumping per unit.

Step 4. Calculate the upper boundary of pumped water:

$$\overline{Q^t_{j,p}} = Min[\overline{Q_{j,p}},\ (V^{t-1}_{j,l} - \underline{V_{j,l}})] \tag{18}$$

If the total volume of pumped water exceed the upper boundary, then decrease the number of pumping units until the upper boundary is satisfied.

Step 5. Calculate the MW power for pumping:

$$P^t_{hj} = -\left(P_{j,sp} \times N^t_p\right) \tag{19}$$

where $P_{j,sp}$ is the constant power for pumping per unit. Then go to step 10.

Step 6. Decode the remaining four bits of the sub-string to obtain a normalized discharge, q^t_j, in decimal values:

Hour t

1	b_2	b_3	b_4	b_5
	×	×	×	×
	2^{-1}	2^{-2}	2^{-3}	2^{-4}

$$q^t_j = \sum_{i=2}^{5}\left(b_i \times 2^{-(i-1)}\right) \qquad b_i \in \{0, 1\} \tag{20}$$

Step 7. Calculate the upper boundary of discharge:

$$\overline{Q^t_j} = Min[\overline{Q_j},\ (\overline{V_{j,l}} - V^{t-1}_{j,l})] \tag{21}$$

Step 8. Translate the normalized value, q^t_j, to the actual value, Q^t_j:

$$Q^t_j = \underline{Q_j} + q^t_j\left(\overline{Q^t_j} - \underline{Q_j}\right) \tag{22}$$

Step 9. Calculate the generation output, P^t_{hj}, using (2).

Step 10. Calculate the remaining thermal loads, P^t_{rm}:

$$P^t_{rm} = P^t_L - P^t_{hj} \tag{23}$$

Step 11. Continue with computations of the 10 steps from hour 1 to hour 24.

Step 12. Perform the unit commitment (UC) for the remaining thermal load profile, and return the corresponding thermal cost. In this work, a UC package based on the neural network (Chen & Chen, 2006) is used to perform the UC task taking into account fuel costs, start-up costs, ramp rate limits, and minimal uptime/downtime constraints.

Step 13. Translate the corresponding thermal cost into the score of the *i*-th particle using a scoring function (details are found in the next Section).

Step 14. Repeat these 13 steps for each particle from the first to last particle.

3.4 Scoring function

The scoring function adopted is based on the corresponding thermal production cost. To emphasize the "best" particles and speed up convergence of the evolutionary process, the scoring function is normalized into a range of 0–1. The scoring function for the *i*-th particle in the swarm is defined as

$$SCORE(i) = \frac{1}{1 + k_i\left(\dfrac{cost(i)}{cost(Gbest)} - 1\right)} \tag{24}$$

where *SCORE(i)* is the score (fitness value) of the *i*-th particle; *cost(i)* is the corresponding thermal cost of the *i*-th particle; *cost(Gbest)* is the cost of the highest ranking particle string, namely, the current best particle; and, k_i is a scaling constant (k_i =100 in this study).

3.5 Mutation operation

The basic PSO approach typically converges rapidly during the initial search period and then slows. Then, checking the positions of particles showed that the particles were very tightly clustered around a local optimum, and the particle velocities were almost zero. This phenomenon resulted in a slow convergence and trapped the whole swarm at a local optimum. Mutation operation is capable of overcoming this shortcoming. Mutation operation is an occasional (with a small probability) random alternation of the *Gbest* string, as shown in Fig. 5. This work integrates a PSO technique with a mutation operation providing background variation and occasionally introduces beneficial materials into the swarm to speed up convergence and escape local optimums.

Gbest:	1	1	1	1	1	0	0	.....	0	1	0
			↓								
New Particle:	1	1	0	1	1	0	0	.....	0	1	0

Figure 5. Mutation operation

The solution methodology for solving the pumped hydro scheduling problem using the proposed approach is outlined in the general flow chart (Fig. 6).

4. Test Results

The proposed approach was implemented on a MATLAB software and executed on a Pentium IV 3.0GHz personal computer. Then, the proposed approach was tested for a portion of the Taipower system, which consists of 34 thermal units and the Ming-Hu pumped hydro plant with four units. In addition to the typical constraints listed in Section 2, the Taipower system has three additional features that increase problem difficulty.

a. The Taipower system is an isolated system. Thus it is self-sufficient at all times. The 300MW system's spinning reserve requirement must be satisfied each hour.
b. Thermal units, due to their ramp rate limits, have difficulty handling large load fluctuations, especially at noon lunch break.
c. The lower reservoir of Ming-Hu pumped hydro plant has only a small storage volume.

Table 1 presents detailed data for the Ming-Hu pumped hydro plant. The thermal system consists of 34 thermal units: six large coal-fired units, eight small coal-fired units, seven oil-fired units, ten gas turbine units, and three combined cycle units. For data on the characteristics of the 34-unit thermal system, please refer to (Chen & Chang, 1995).

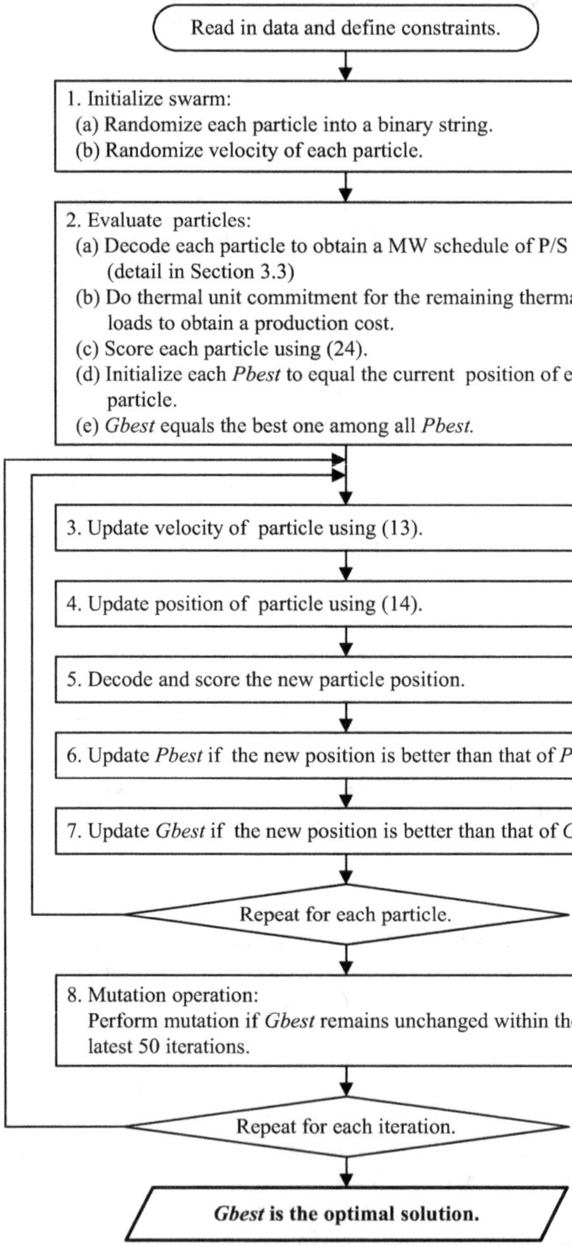

Figure 6. General flow chart of the proposed approach

			Lower Reservoir		
Installed Capacity	Maximal Discharge (m^3/s)	Maximal Pumping (m^3/s)	Maximal Storage ($10^3\,m^3$)	Minimal Storage ($10^3\,m^3$)	Efficiency
250MW×4	380	249	9,756	1,478	≈0.74

Table 1. Characteristics of the Ming-Hu pumped hydro plant

The proposed approach was tested on a summer weekday whose load profile (Fig. 7) was obtained by subtracting expected generation output of other hydro plants and nuclear units from the actual system load profile. Fig. 8 and 9 present schematics of test results. Fig. 8 shows the total generation/pumping schedules created by the proposed approach. Fig. 9 shows the remaining thermal load profiles. The optimal schedules for pumped hydro units and thermal units are obtained within 3 minutes, satisfying Taipower's time requirement.

To investigate further how the proposed approach and existing methods differ in performance, this work adopts a DP method (Chen, 1989) and a GA method (Chen & Chang, 1999) as the benchmark for comparison. Table 2 summarizes the test results obtained using these three methods.

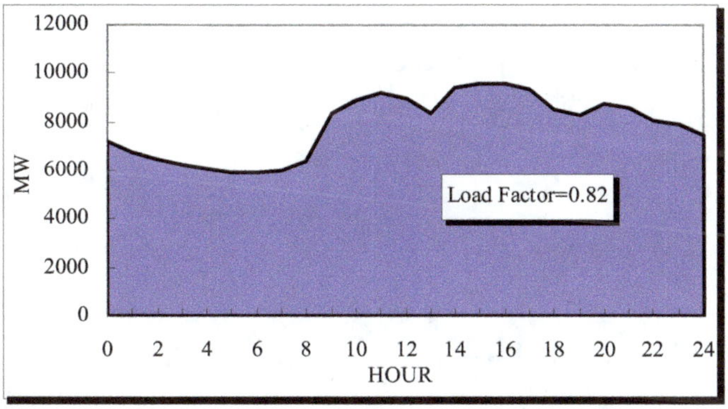

Figure 7. Summer weekday load profile

Several interesting and important observations are derived from this study and are summarized as follows:

a. The generation/pumping profile generally follows the load fluctuation, a finding that is consistent with economic expectations. The Ming-Hu pumped hydro plant generates 3,893 MWh power during peak load hours and pumps up 5,250 MWh power during light load hours, resulting in a cost saving of NT$5.91 million in one day, where cost saving = (cost without pumped hydro) - (cost with pumped hydro).
b. The pumped hydro units are the primary source of system spinning reserve due to their fast response characteristics. The system's spinning reserve requirement accounts for the fact that pumped hydro units do not generate power at their maximum during peak load hours.
c. Variation of water storage in the small lower reservoir is always retained within the maximum and minimum boundaries. The final volume returns to the same as the initial volume.
d. The load factor is improved from 0.82 to 0.88 due to the contribution of the four pumped hydro units.

e. Notably, both cost saving and execution time for the proposed approach are superior to either a DP or a GA method.

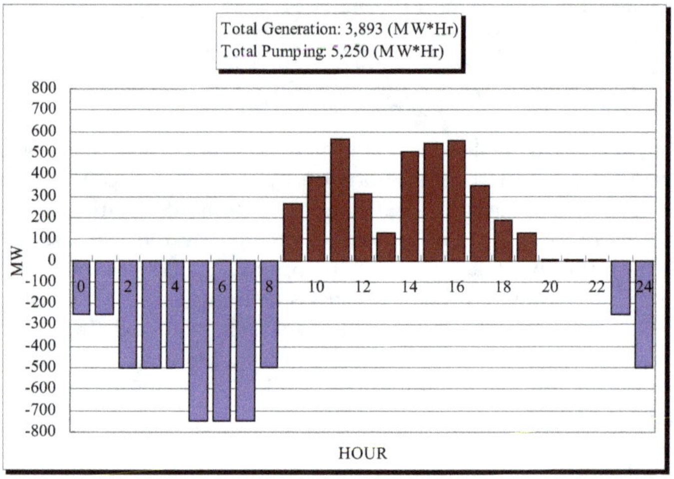

Figure 8. Hourly generation/pumping schedules

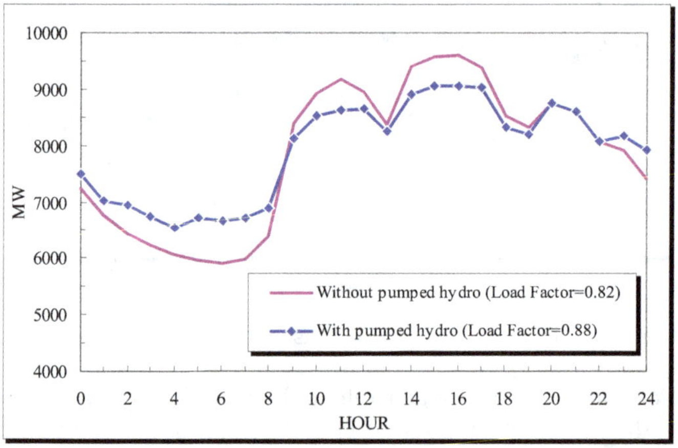

Figure 9. Contrast of two remaining thermal load profiles

Method	Load Factor	Cost Saving (10^3 NT$)	Execution Time (second)
DP	0.87	5,641	336
GA	0.87	5,738	164
RPSO	0.88	5,906	127

Table 2. Performance comparison with existing methods

5. Conclusion

This work presents a novel methodology based on a refined PSO approach for solving the power dispatch with pumped hydro problem. An advantage of the proposed technique is the flexibility of PSO for modeling various constraints. The difficult water dynamic balance constraints are embedded and satisfied throughout the proposed encoding/decoding algorithms. The effect of net head, constant power pumping characteristic, thermal ramp rate limits, minimal uptime/downtime constraints, and system's spinning reserve requirements are all considered in this work to make the scheduling more practical. Numerical results for an actual utility system indicate that the proposed approach has highly attractive properties, a highly optimal solution and robust convergence behavior for practical applications.

6. References

Al-Agtash, S. (2001). Hydrothermal scheduling by augmented Lagrangian: consideration of transmission constraints and pumped-storage units, *IEEE Transactions on Power System*, Vol. 16, pp. 750-756

Allan, R. N. & Roman, J. (1991). Reliability assessment of hydrothermal generation systems containing pumped storage plant, *IEE Proceedings-C, Generation, Transmission, and Distribution,* Vol. 138, pp. 471-478

Angeline, P. J. (1998). Evolutionary optimization versus particle swarm optimization: philosophy and performance differences, *Lecture Notes in Computer Science,* Vol. 1447, pp. 601–610

Boeringer, D. W. & Werner, D. H. (2004). Particle swarm optimization versus genetic algorithms for phased array synthesis, *IEEE Transactions on Antennas Propagation*, Vol. 52, pp. 771-779

Chen, P. H. (1989). Generation scheduling of hydraulically coupled plants, *Master's thesis,* Dept. Elect. Eng., National Tsing-Hua University, Taiwan

Chen, P. H. & Chang, H. C. (1995). Large-scale economic dispatch by genetic algorithm, *IEEE Transactions on Power System,* Vol. 10, pp. 1919-1926

Chen, P. H. & Chang, H. C. (1999). Pumped-storage scheduling using a genetic algorithm, *Proceedings of the Third IASTED International Conference on Power and Energy Systems,* pp. 492-497, USA

Chen, P. H. & Chen, H. C. (2006). Application of evolutionary neural network to power system unit commitment, *Lecture Notes in Computer Science,* Vol. 3972, pp. 1296–1303

Eberhart, R. & Shi, Y. (1998). Comparison between genetic algorithms and particle swarm optimization, *Lecture Notes in Computer Science,* vol. 1447, pp. 611–616

Eberhart, R. & Shi, Y. (2001). Particle swarm optimization: developments, applications and resources, *Proceedings of IEEE International Congress on Evolutionary Computation,* Vol. 1, pp. 81–86

El-Hawary, M. E. & Ravindranath, K. M. (1992). Hydro-thermal power flow scheduling accounting for head variations, *IEEE Transactions on Power System,* Vol. 7, pp. 1232-1238

Guan, X.; Luh, P. B.; Yen, H. & Rogan, P. (1994). Optimization-based scheduling of hydrothermal power systems with pumped-storage units, *IEEE Transactions on Power System*, Vol. 9, pp. 1023-1029

Ho, S. L.; Yang, S.; Ni, G. & Wong, H. C. (2006). A particle swarm optimization method with enhanced global search ability for design optimizations of electromagnetic devices, *IEEE Transactions on Magnetics*, Vol. 42, pp. 1107-1110

Jeng, L. H.; Hsu, Y. Y.; Chang, B. S. & Chen, K. K. (1996). A linear programming method for the scheduling of pumped-storage units with oscillatory stability constraints, *IEEE Transactions on Power System*, Vol. 11, pp. 1705-1710

Juang, C. F. (2004). A hybrid of genetic algorithm and particle swarm optimization for recurrent network design, *IEEE Transactions on Systems, Man, and Cybernetics*, Vol. 34, pp. 997-1006

Kennedy, J. & Eberhart, R. (1995). Particle swarm optimization, *Proceedings of IEEE International Conference on Neural Networks*, Vol. 4, pp. 1942–1948

Kennedy, J. & Eberhart, R. (1997). A discrete binary version of the particle swarm algorithm, *Proceedings of International Conference on Systems, Man, and Cybernetics*, pp. 4104–4109, Piscataway, NJ

Kennedy, J. & Eberhart, R. (2001). *Swarm Intelligence*, Morgan Kaufmann, San Francisco

Kuo, C. C.; Chen, P. H.; Hsu, C. L. & Lin, S. S. (2005). Particle swarm optimization for capacitor allocation and dispatching problem using interactive best-compromise approach, *WSEAS Transactions on System*, Vol. 4, pp. 1563-1572

Naka, S.; Genji, T.; Yura, T. & Fukuyama, Y. (2003). A hybrid particle swarm optimization for distribution state estimation, *IEEE Transactions on Power System*, Vol. 18, pp. 60-68

Wood, A. J. & Wollenberg, B. F. (1996). *Power Generation, Operation, and Control*, 2nd ed., John Wiley & Sons, New York

Zhao, B.; Guo, C. X. & Cao, Y. J. (2005). A multiagent-based particle swarm optimization approach for optimal reactive power dispatch, *IEEE Transactions on Power System*, Vol. 20, pp. 1070-1078

10

Particle Swarm Optimization Applied to Parameters Learning of Probabilistic Neural Networks for Classification of Economic Activities

Patrick Marques Ciarelli, Renato A. Krohling and Elias Oliveira
Universidade Federal do Espírito Santo
Brazil

1. Introduction

Automatic text classification and clustering are still very challenging computational problems to the information retrieval (IR) communities both in academic and industrial contexts. Currently, a great effort of work on IR, one can find in the literature, is focused on classification and clustering of generic content of text documents. However, there are many other important applications to which little attention has hitherto been paid, which are as well very difficult to deal with. One example of these applications is the classification of companies based on the descriptions of their economic activities, also called mission statements, which represent the business context of the companies' activities, in other words, the business economic activities from free text description by the company's founders.

The categorization of companies according to their economic activities constitute a very important step towards building tools for obtaining correct information for performing statistical analysis of the economic activities within a city or country. With this goal, the Brazilian government is creating a centralized digital library with the business economic activity descriptions of all companies in the country. This library will serve the three government levels: Federal; the 27 States; and more than 5.000 Brazilian counties. We estimate that the data related to nearly 1.5 million companies will have to be processed every year (DNRC, 2007) into more than 1.000 possible different activities. It is important to highlight that the large number of possible categories makes this problem particularly complex when compared with others presented in the literature (Jain et al., 1999; Sebastiani, 2002).

In this paper, we proposed a slightly modified version of the standard structure of the probabilistic neural network (PNN) (Specht, 1990) so that we could deal with the multi-label problem faced in this work. We compared the PNN performance trained by a canonical Particle Swarm Optimization (PSO) and a Bare Bones Particle Swarm Optimization (BBPSO). Our results show that, in the categorization of free text descriptions of economic activities, the PNN trained by BBPSO got slightly better results than the PNN trained by PSO.

This work is organized as follows. In Section 2, we detail more the characteristics of the problem and its importance for the government institutions in Brazil. Related works are mentioned in Section 3. We describe our probabilistic neural network algorithm in Section 4. Section 5 describes the Particle Swarm Optimization algorithm and a special version named Bare Bones Particle Swarm Optimization. In Section 6, the experimental results are discussed. Finally, we present our conclusions and indicate some future paths for future research in Section 7.

2. The Problem of Multi-label Text Categorization

In many countries, companies must have a contract (Articles of Incorporation or Corporate Charter, in USA) with the society where they can legally operate. In Brazil, this contract is called a social contract and must contain the statement of purpose of the company - this statement of purpose describe the business activities of the company and must be categorized into a legal business activity by Brazilian government officials. For that, all legal business activities are cataloged using a table called National Classification of Economic Activities, for short, CNAE (CNAE, 2003).

To perform the categorization, the government officials (at the Federal, State and County levels) must find the semantic correspondence between the company economic activities description and one or more entries of the CNAE table. There is a numerical code for each entry of the CNAE table and, in the categorization task, the government official attributes one or more of such codes to the company at hand. This can happen on the foundation of the company or in a change of its social contract, if that modifies its economic activities.

The work of finding the semantic correspondence between the company economic activities description and a set of entries into the CNAE table are both very difficult and labor-intensive task. This is because of the subjectivity of each local government officials who can focus on their own particular interests so that some codes may be assigned to a company, whereas in other regions, similar companies, may have a totally different set of codes. Sometimes, even inside of the same state, different level of government officials may count on a different number of codes for the same company for performing their work of assessing that company. Having inhomogeneous ways of classifying any company everywhere in all the three levels of the governmental administrations can cause a serious distortion on the key information for the long time planning and taxation. Additionally, the continental size of Brazil makes this problem of classification even worse.

In addition, the number of codes assigned by the human specialist to a company can vary greatly, in our dataset we have seen cases where the number of codes varied from 1 up to 109. However, in the set of assigned codes, the first code is the main code of that company. The remaining codes have no order of importance.

Due to this task is up to now decentralized, we might have the same job being performed many times by each of the three levels of the government officials. Nevertheless, it is known that there has been not enough staff to do this job properly.

For all these reasons, the computational problem addressed by us is mainly that of automatically suggesting the human classifier the semantic correspondence between a textual description of the economic activities of a company and one or more items of the CNAE table. Or, depending on the level of certainty the algorithms have on the automatic classification, we may consider bypassing thus the human classifier.

2.1 Metrics for Evaluating of Multi-label Text Categorization

Typically, text categorization is mainly evaluated by the Recall and Precision metrics in the single-labled cases (Baeza-Yates & Ribeiro-Neto, 1998). Nonetheless, other authors have already proposed different metrics for multi-label categorization problems (Schapire & Singer, 2000; Zhang & Zhou, 2007).

Formalizing the problem we have at hand, text categorization may be defined as a task of assigning documents to a predefined set of categories, or classes (Sebastiani, 2002). In multi-label text categorization a document may be assigned to one or more categories. Let D be the domain of documents, $C = \{c_1, c_2, \ldots, c_{|C|}\}$ a set of predefined categories, and $W = \{d_1, d_2, \ldots, d_{|W|}\}$ an initial set of documents previously categorized by some human specialists into subsets of categories of C.
In multi-label learning, the training (-and validation) set $TV = \{d_1, d_2, \ldots, d_{|TV|}\}$ is composed of a number of documents, each associated with a subset of categories in C. TV is used to train and validate (actually, to tune eventual parameters of) a categorization system that associates the appropriate combination of categories to the characteristics of each document in the TV. The test set $Te = \{d_{|TV|+1}, d_{|TV|+2}, \ldots, d_{|W|}\}$, on the other hand, consists of documents for which the categories are unknown to the automatic categorization systems. After being trained, as well as tuned, by the TV, the categorization systems are used to predict the set of categories of each document in Te.

A multi-label categorization system typically implements a real-valued function of the form $f : D \times C \rightarrow \Re$ that returns a value for each pair $\langle d_j, c_i \rangle \in D \times C$ that, roughly speaking, represents the evidence for the fact that the test document d_j should be categorized under the category $c_i \in C_i$, where $C_i \subset C$. The real-valued function $f(.,.)$ can be transformed into a ranking function $r(.,.)$, which is an one-to-one mapping onto $\{1, 2, \ldots, |C|\}$ such that, if $f(d_j, c_1) > f(d_j, c_2)$, then $r(d_j, c_1) < r(d_j, c_2)$. If C_i is the set of proper categories for the test document d_j, then a successful categorization system tends to rank categories in C_i higher than those not in C_i. Additionally, we also use a threshold parameter so that those categories that are ranked above the threshold τ (i.e., $c_k \mid f(d_j, c_k) \geq \tau$) are the only ones to be assigned to the test document.

We have used five multi-label metrics discussed by Zhang & Zhou (2007) to evaluate the categorization performance of PNN: *hamming loss, one-error, coverage, ranking loss,* and *average precision*. We now present each of these metrics:

- **Hamming Loss (hloss$_j$)** evaluates how many times the test document d_j is misclassified, i.e., a category not belonging to the document is predicted or a category belonging to the document is not predicted.

$$\text{hloss}_j = \frac{1}{|C|} |P_j \Delta C_i| \tag{1}$$

where $|C|$ is the number of categories and Δ is the symmetric difference between the set of predicted categories P_j and the set of appropriate categories C_i of the test document d_j. The predicted categories are those with rank higher than the threshold τ.

- **One-error (one-error$_j$)** evaluates if the top ranked category is present in the set of proper categories C_i of the test document d_j.

$$\text{one-error}_j = \begin{cases} 0 & \text{if} \quad \arg\max_{c \in C} f(d_j, c) \in C_i \\ 1 & \text{otherwise} \end{cases} \tag{2}$$

where $\arg\max_{c \in C} f(d_j, c)$ returns the top ranked category for the test document d_j.

- **Coverage (coverage$_j$)** measures how far we need to go down the rank of categories in order to cover all the possible categories assigned to a test document.

$$\text{coverage}_j = \max_{c \in C_i} r(d_j, c) - 1 \tag{3}$$

where $\max_{c \in C_i} r(d_j, c)$ returns the maximum rank for the set of appropriate categories of the test document d_j.

- **Ranking Loss (rloss$_j$)** evaluates the fraction of category pairs $\langle c_k, c_l \rangle$, for which $c_k \in C_i$ and $c_l \in \overline{C_i}$, that are reversely ordered for the test document d_j:

$$\text{rloss}_j = \frac{\left|\{(c_k, c_l) | f(d_j, c_k) \le f(d_j, c_l)\}\right|}{|C_i||\overline{C_i}|} \tag{4}$$

where $(c_k, c_l) \in C_i \times \overline{C_i}$, and $\overline{C_i}$ is the complementary set of C_i in C.

- **Average Precision (avgprec$_j$)** evaluates the average of precisions computed after truncating the ranking of categories after each category $c_i \in C_i$ in turn:

$$\text{avgprec}_j = \frac{1}{|C_i|} \sum_{k=1}^{|C_i|} \text{precision}_j(R_{jk}) \tag{5}$$

where R_{jk} is the set of ranked categories that goes from the top ranked category until a ranking position k where there is a category $c_i \in C_i$ for d_j, and $\text{precision}_j(R_{jk})$ is the number of pertinent categories in R_{jk} divided by $|R_{jk}|$.

For p test documents, the overall performance is obtained by averaging each metric, that is,

$$\text{hloss} = \frac{1}{p}\sum_{j=1}^{p} \text{hloss}_j, \qquad \text{one-error} = \frac{1}{p}\sum_{j=1}^{p} \text{one-error}_j, \qquad \text{coverage} = \frac{1}{p}\sum_{j=1}^{p} \text{coverage}_j,$$

$\text{rloss} = \frac{1}{p}\sum_{j=1}^{p} \text{rloss}_j$, $\text{avgprec} = \frac{1}{p}\sum_{j=1}^{p} \text{avgprec}_j$. On the one hand, the smaller the value of *hamming loss, one-error, coverage* and *ranking loss,* the better the performance of the categorization system. On the other hand, for the *average precision*, the larger the value the better the performance. So, the problem can be formulated as an optimization problem, where the performance is optimal when hloss = one-error = rloss = 0 and avgprec = 1.

In the next section are mentioned some related works regarding the problem of economic activities classification.

3. Related Works

The authors in (Souza et al., 2007) are among the first to tackle the problem of economic activities classification. In their work they compared the results achieved between a Nearly Neighbors algorithm approach and a Weightless Neural Network, called VG-RAM WNN, using a metric to evaluate the performance equivalent to 1 - one-error, defined in Section 2.1. In the first algorithm they got the performance of 63.36%, while VG-RAM WNN showed to be slightly better, with a performance of 67.56%. However, the use of a single metric seemed to be not enough for evaluating multi-labled problems.

A different approach was performed by (Oliveira et al., 2007). In this work were used 83 arrays of small standard PNN for classification, whose main metrics used were Recall and Precision. However, it was noted to be very difficult to merge the results returned of each neural network array node. Thus the performance of the array as a whole was harmed. Although it has found a reasonable value for the Recall, the value for the Precision was very low, since almost every neural networks returned at least one class to each instance of test.

A PNN with a slightly modified architecture to treat problems of multi-label classification was proposed in (Oliveira et al., 2008). Such neural network presents advantage over the array of small standard PNN approach, used in (Oliveira et al., 2007), because only one PNN is used to solve the problem of multi-label classification. Whereas, in the previous approach, we need to build many neural networks (83 in that case) which complicate the process of optimization.

The results achieved in (Oliveira et al., 2008) using the proposed PNN were better than the achieved using the Multi-label k-Nearest Neighbors (ML-kNN) algorithm. The ML-kNN was considered to be the best algorithm for all the database used in (Zhang & Zhou, 2007). In order to evaluate the performance of the algorithms, the authors in that work used the metrics presented in the Section 2. Moreover, the parameters of these algorithms were optimized using a Genetic Algorithm (GA).

The cited previous works used the same database that we present in this work, but the division of the database was performed in a different way for each work, making it difficult conducting a comparison of results among them. However, in this work we will divide the database in a similar way to used in (Oliveira et al., 2008), making possible a comparison among results.

Another very close multi-label problem to one we are presenting in this paper, concern with the economic activities classification, is that of patent categorization (Li et al., 2007). Our problem and that are both based on free text descriptions of variety topics. So a large volume of patents documents, are usually, up to these days, manually classified by the patent offices, this is a labor-intensive and time-consuming task. A patent document may cite another patent document, or articles, for comparing or contrasting reasons. Therefore, besides using the content categorization approach, the authors in (Li et al., 2007) proposed to extract and use the direct hyperlink citation relationships among patent documents in order to improve the quality of the whole process of classification. Hyperlink citation is a similar strategy some researchers have been widely applied to web page classification studies. The experiments were conducted on a nanotechnology-related patent dataset from the USPTO. The training dataset contained 13,913 instances, and the testing data set 4,358 data instances. The average of category for document was 36, and the total of categories was up to 426. The results by the K_{Gra} kernel proposed approach yielded 86.67% accuracy overcome the 81% of manually processing and the results of previous work (Koster et al., 2003).

In the following, we describe a slightly modified Probabilistic Neural Network (PNN) used to solve the optimization problem of text categorization.

4. Probabilistic Neural Network Architecture

The Probabilistic Neural Network was first proposed by Donald Specht in 1990 (Specht, 1990). This is an artificial neural network for nonlinear computing, which approaches the Bayes optimal decision boundaries. This is done by estimating the *probability density function* of the training dataset using the Parzen nonparametric estimator (Parzen, 1962).

The literature has shown that this type of neural network can yield similar results, sometimes superior, in pattern recognition problems when compared with others techniques (Fung et al., 2005; Patra et al., 2002).

The original Probabilistic Neural Network algorithm was designed for single-label problems. Thus, we slightly modified its standard architecture, so that it is now capable of solving multi-label problem addressed in this work.

In our modified version, instead of four, the Probabilistic Neural Network is now composed of only three layers: the *input* layer, the *pattern* layer and the *summation* layer, as depicted in Figure 1. Thus like the original, this version of Probabilistic Neural Network needs only one training step, thus its training is very fast compared to the others feedforward neural networks (Duda et al., 2001; Haykin, 1998). The training consists in assigning each training sample w_i of class C_i to a neuron of pattern layer of class C_i. Thus the weight vector of this neuron is the characteristics vector of the sample.

For each pattern x passed by the input layer to a neuron in the pattern layer, it computes the output for x. The computation is performed by Equation 6.

$$F_{k,i}(x) = \frac{1}{2\pi\sigma^2}\exp\left(\frac{x^t w_{ki} - 1}{\sigma^2}\right) \tag{6}$$

where x is the pattern characteristics input vector, and the w_{ki} is the kth sample for a neuron of class C_i, $k \in N_i$, whereas N_i is the number of neurons of C_i. In addition, x was normalized so that $x'x = 1$ and $w_{ki}{}'w_{ki} = 1$. The parameter σ is the Gaussian standard deviation, which determines the receptive field of the Gaussian curve.

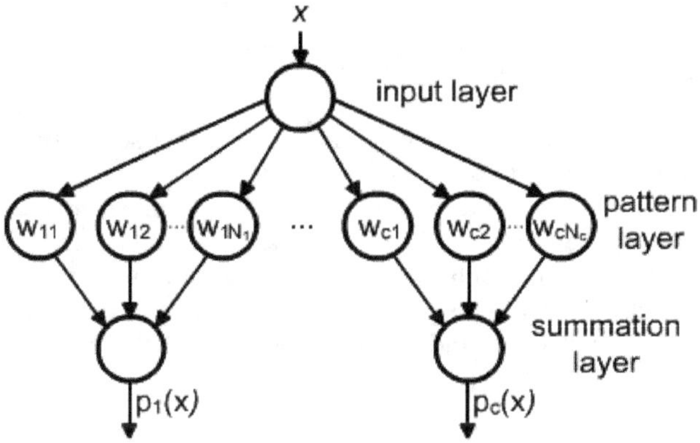

Figure 1. The modified Probabilistic Neural Network architecture

The next step is the summation layer. In this layer, all weight vectors are summed according to Equation 7, in each cluster C_i producing $p_i(x)$ values, where $|C|$ is the total number of classes.

$$p_i(x) = \sum_{k=1}^{N_i} F_{k,i}(x) \tag{7}$$

$$k = 1,2,\ldots,N_i; \quad i = 1,2,\ldots,|C|$$

Finally, for the selection of the classes, which will be assigned by neural network to each sample, we consider the most likely classes pointed out by the summation layer based on a chosen threshold.

Differently from other types of neural networks, such as the feedforward one (Haykin, 1998), the probabilistic neural network proposed needs few parameters to be configured: the σ, (see Equation 6) and the determination of threshold value. The σ is used to narrow the receptive field of the Gaussian curve in order to strictly select only the more likely inputs for a given class. Other advantages of the probabilistic neural networks is that it is easy to add new classes, or new training inputs, into the already running structure, which is good for on-line applications (Duda et al., 2001). Moreover, it is reported in the literature (Duda et al., 2001) that it is also easy to implement this type of neural network in parallel. On the other hand, one of its drawbacks is the great number of neurons in the pattern layer, which can be, nevertheless, mitigated by an optimization on the number of the neuron (Georgiou et al., 2004; Mao et al., 2000).

Next, we propose a PSO algorithm to find out the σ parameters and tune the PNN automatically.

5. The Canonical and the Bare Bones Particle Swarm Optimization

Particle Swarm Optimisation (PSO) has its origins in the simulation of bird flocking developed by Reynolds (1987) and was further developed in the context of optimization by Eberhart and Kennedy (Eberhart & Kennedy, 1995; Kennedy & Eberhart, 1995). PSO is initialised with a population of random solutions. Each potential solution in PSO is also

associated with a randomised velocity, and the potential solutions, are called *particles*, that move in the search space. Each particle keeps track of its coordinates in the problem space, which are associated with the best solution (fitness) it has achieved so far. This value is called *pbest*. Another *best* value that is tracked by the *global* version of the particle swarm optimizer is the overall best value, and its location, obtained so far by any particle in the population. This location is called *gbest*.
The particle swarm optimization concept consists of, at each time step, changing the velocity of each particle moving toward its *pbest* and *gbest* locations (global version of PSO). Acceleration is weighted by random terms, with separate random numbers being generated for acceleration toward *pbest* and *gbest* locations, respectively. The PSO algorithm consists basically in updating the velocities and positions of the particle, respectively as follows in Equations 8 and 9 (Clerc & Kennedy, 2002):

$$v_i(t+1) = \lambda\left[v_i(t) + c_1 rand_1\left(p_{best_i} - x_i(t)\right) + c_2 rand_2\left(g_{best} - x_i(t)\right)\right] \tag{8}$$

$$x_i(t+1) = x_i(t) + v_i(t+1) \tag{9}$$

$$\text{with} \quad \lambda = \frac{2}{\left|\sqrt{2-\varphi-\sqrt{\varphi^2-4\varphi}}\right|}, \quad \text{where} \quad \varphi = c_1 + c_2, \quad \varphi > 4$$

where:

- $x_i = [x_{i1}, x_{i2}, \ldots, x_{in}]^T$ is the position of the ith particle in the n-dimensional search space;
- $v_i = [v_{i1}, v_{i2}, \ldots, v_{in}]^T$ is the velocity of the ith particle;
- p_{best_i} is the best previous i^{th} particle position;
- g_{best} is the best particle among all particles;
- λ is the constriction factor;
- c_1 and c_2 are positive constants;
- $rand_1$ and $rand_2$ are random numbers in the range [0;1] generated using the uniform probability distribution.

Usually, when the constriction factor is used, φ is set to 4.1 ($c_1 = c_2 = 2.05$), and the constriction factor λ is 0.729. In this paper, it is assumed minimization problems unless stated otherwise.

In the meantime different versions of PSO have been proposed by (Krohling & Coelho, 2006). In this work we focus on the Bare Bones PSO (Kennedy, 2003). The Bare Bones PSO (BBPSO) eliminates the velocity item and the Gaussian distribution is used to sampling the search space based on the global best (gbest) and the personal best (pbest) particle. So, the Equations 8 and 9 are replaced by Equation 10:

$$N = \left(\mu_i, \sigma_i^2\right) \tag{10}$$

$$\text{with} \quad \mu_i = \frac{\left(g_{best} + p_{best_i}\right)}{2}, \quad \sigma_i = \left|g_{best} - p_{best_i}\right|$$

where N denotes the Gaussian distribution.

This version of PSO presents some advantages over other versions because its reduced numbers of parameters of the algorithms to be tuned. The BBPSO is described in the Listing 1.

```
PSO and BBPSO Algorithms
Input parameters: swarm size P
FOR each particle i
// random initialization of a population of particles with positions x_i using uniform
// probability distribution.
  x_i = \underline{x_i} + (\overline{x_i} - \underline{x_i})·u_i  // \underline{x_i} and \overline{x_i} stands for the lower and upper bound,
//respectively, and u_i is a random number.
  p_best_i = x_i
  compute f(x_i) // fitness evaluation.
  p_gbest := arg min{f(x_i)} // global best particle.
END FOR
DO
  FOR each particle i
    update the position x_i according to Equations 8 and 9 if PSO
    update the position x_i according to Equation 10 if BBPSO
    compute f(x_i) // fitness evaluation
    IF f(x_i) < f(p_best_i) THEN // update of the personal best.
      p_best_i = x_i
    IF f(x_i) < f(p_gbest) THEN // update the global best.
      p_gbest = p_best_i
  END FOR
WHILE termination condition not met.
Output: p_gbest, f(p_gbest).
```

Listing 1 PSO and BBPSO Algorithms.

6. Experimental Results

We employed a series of experiments to compare PNNs optimized using canonical PSO and BBPSO. We used a dataset containing 3264 documents of free text business descriptions of Brazilian companies categorized into a subset of 764 CNAE categories. This dataset was obtained from real companies placed in Vitoria County in Brazil. The CNAE codes of each company in this dataset were assigned by Brazilian government officials trained for this task. Then we evenly partitioned the whole dataset into four subsets of equal size of 816 documents. We joined to this categorizing dataset the brief description of each one of the 764 CNAE categories, totalizing 4028 documents. Hence, in all training (-and validation) set, we adopted the 764 descriptions of CNAE categories and a subset of 816 business description documents, and, as the test set, the other three subsets of business descriptions totalizing 2448 documents.

6.1 Categorization of Free-text Descriptions of Economic Activities

We pre-processed the dataset via term selection - a total of 1001 terms were found in the database after removing stop words and trivial cases of gender and plural; only words appearing in the CNAE table were considered. After that, each document in the dataset was described as a multidimensional vector using the Bag-of-Words representation (Dumais et al., 1998), i.e., each dimension of the vector corresponds to the number of times a term of the vocabulary appears in the corresponding document. Table 1 summarizes the characteristics of this dataset (dataset available at http://www.inf.ufes.br/~elias/vitoria.tar.gz).

#C	#t	Training set				Test/validation set			
		NTD	DC	CD	RC	NTD	DC	CD	RC
764	1001	4.65	0.00	1.00	100.00	10.92	74.48	4.27	85.21

Table 1. Characteristics of the CNAE dataset

In this Table #C denotes the number of categories, #t denotes the number of terms in the vocabulary, NTD denotes the average number of terms per document, DC denotes the percentage of documents belonging to more than one category, CD denotes the average number of categories for each document, and RC denotes the percentage of rare categories, i.e., those categories associated with less than 1% of the documents of the dataset. The training set is composed by 764 categories descriptions belonging at CNAE table, where each description is concerning just one category and there is only one description by category (one to one relationship), resulting in CD equal 1 and DC equal 0. As there are 764 instances of training and just one instance for category, the index RC is equal 100%. On the other hand, the test/validation set is composed by 3264 instances, where 74.48% of instances are assigned to more than one category and the average number of categories of each instance is more than 4 per document. However, like we said in Section 2, this number vary greatly. Moreover, we can note that RC value is high since there are few instances by category.

The PNNs parameters σ, in Equation 6, were optimized for each class of the dataset and just one threshold τ value for the whole neural network, resulting in 765 parameters, i. e., each particle is represented by a 765-dimensional vector. This is a quite huge amount of parameters for optimization.

To tune these parameters we divided the training set (-and validation) set into a training set, which was used to inductively build the categorizer, and a validation set, which was used to evaluate the performance of the categorizer in the series of experiments aimed at parameter optimization. The training set is composed of 764 descriptions of CNAE classes and the validation set of 816 business description documents described previously. As a result, we carried out a sequence of experiments with PSO and BBPSO. For each one of these algorithms was carried out 48 experiments:

- 4 experiments each using algorithm with 100 particles and 500 iterations;
- 4 experiments each using algorithm with 50 particles and 500 iterations;
- 40 experiments each using algorithm with 50 particles and 100 iterations.

The two first experiments set were used to evaluate the performance of the algorithms for different population sizes. The last 40 experiments were used for a statistical analysis.
In Figures 2 and 3 are shown the performance of the PNN optimized in function of the number of iterations for 100 and 50 particles, respectively. Where is written in the legend 1st subset means that the first subset was used for validation and the 764 descriptions were used for training, in a similar way this is valid for others cases. The continuous lines are the results of the canonical PSO algorithm and the dotted lines are the results of the BBPSO algorithm. Here, the performance value is a linear combination of the several metrics, where these metrics were described in the Section 2. Thus, performance is the sum of the hamming loss, one error, coverage, ranking loss and $\overline{\text{average precision}}$, where $\overline{\text{average precision}}$ = 1 - average precision. The coverage value was divided by the factor $|C|-1$ to normalize it and keep it in the same scale of the others metrics. A strategy for optimization could be the use of weighted metrics, however in this work was regarded the same value of importance for every metrics.

In both figures the smaller the value of the performance, the better the performance of the neural network. We can observe in both figures that the BBPSO algorithm presented better results than the canonical PSO algorithm. Although the determination of the optimal swarm size is beyond the scope of this work, can be noted that exist no big differences between the results obtained with 100 particles and 50 particles. Moreover, there is a large gain of performance until the 100th iteration and a gain slower in the next iterations. Because of this and since the experiments require substantial amount of run time, we carry out others experiments using 50 particles and 100 iterations for statistical analysis purposes.
In the Table 2 are shown the best, mean, median, standard deviation and worst results obtained in the validation with PSO and BBPSO. The results in bold indicate the best results found for each subset. We can observe in Table 2 that BBPSO finds slightly better results than the canonical PSO.

After tuning, the multi-label categorizers were trained with the 764 descriptions of CNAE categories and tested with the 2448 documents of the test set. The Table 3 shows the best, mean, median, standard deviation and worst results found in the validation with PSO and BBPSO. In this table, where is written 1st means the 1st subset for validation and the others subsets for test, in a similar way this is valid for the other subsets. Again, the results in bold are the best results found for each subset. Similarly as occurred in Table 2, Table 3 also shows that the BBPSO performs slightly better than the PSO.

The mean of results achieved for each metric are shown in Table 4 and 5 for the PNN trained by canonical PSO and BBPSO, respectively. Comparing the results found in this tables we noticed that there weren't significant differences among them, this indicates that the proposed PNN presents certain robustness on the dataset used for training/validation.

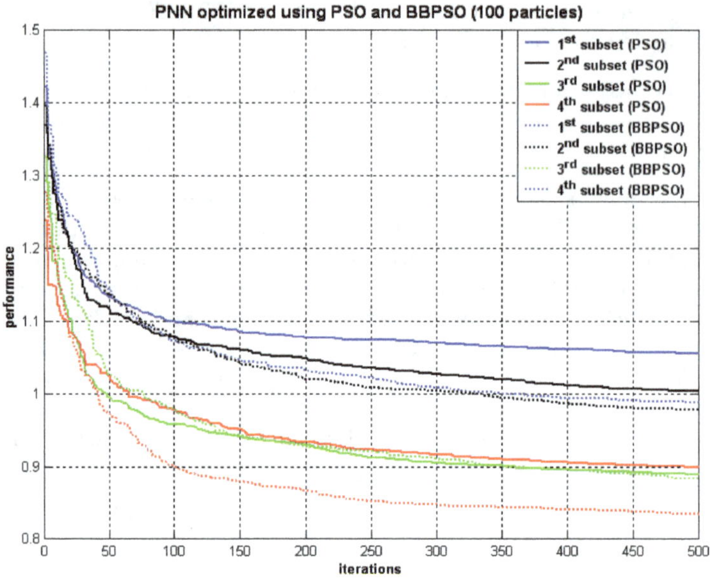

Figure 2. Experimental results of validation of the PNN using PSO and BBPSO with 100 particles and 500 iterations

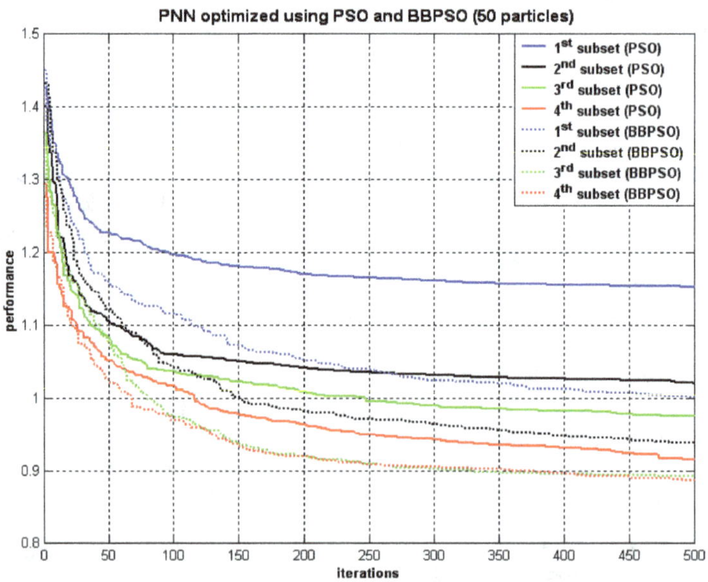

Figure 3. Experimental results of validation of the PNN using PSO and BBPSO with 50 particles and 500 iterations

Subset	Algorithm	Best	Mean	Median	Std. Deviation	Worst
1st	PSO	1.1334	1.1744	1.1617	0.0292	1.2194
	BBPSO	**1.0789**	**1.1107**	**1.1163**	**0.0187**	**1.1344**
2nd	PSO	1.0596	1.0971	1.1003	**0.0176**	1.1241
	BBPSO	**1.0334**	**1.0652**	**1.0671**	0.0192	**1.0864**
3rd	PSO	1.0320	1.0475	1.0488	**0.0087**	1.0608
	BBPSO	**0.9622**	**0.9902**	**0.9842**	0.0217	**1.0257**
4th	PSO	0.9741	1.0060	1.0072	0.0202	1.0435
	BBPSO	**0.9363**	**0.9553**	**0.9558**	**0.0110**	**0.9746**

Table 2. Information about the validation phase

Subset	Algorithm	Best	Mean	Median	Std. Deviation	Worst
1st	PSO	1.1465	1.1766	1.1671	0.0284	1.2332
	BBPSO	**1.1204**	**1.1361**	**1.1358**	**0.0146**	**1.1689**
2nd	PSO	1.1190	1.1632	1.1679	0.0217	1.1853
	BBPSO	**1.1107**	**1.1429**	**1.1432**	**0.0209**	**1.1798**
3rd	PSO	1.1537	1.1873	1.1912	0.0203	1.2202
	BBPSO	**1.1375**	**1.1632**	**1.1633**	**0.0184**	**1.2026**
4th	PSO	1.1841	1.2260	1.2260	0.0345	1.3057
	BBPSO	**1.1555**	**1.1810**	**1.1826**	**0.0167**	**1.2094**

Table 3. Information about the test phase

Subeset	Hamming loss	One-error	Coverage	Ranking loss	Average precision
1st	0.0056	0.3708	144.3723	0.0835	0.4725
2nd	0.0056	0.3688	142.2521	0.0874	0.4850
3rd	0.0055	0.3756	143.9365	0.0912	0.4737
4th	0.0056	0.3857	154.7565	0.0937	0.4619

Table 4. Results achieved with PNN trained by canonical PSO

Subeset	Hamming loss	One-error	Coverage	Ranking loss	Average precision
1st	0.0056	0.3544	143.9079	0.0749	0.4875
2nd	0.0056	0.3592	145.8953	0.0805	0.4937
3rd	0.0055	0.3648	147.5607	0.0842	0.4847
4th	0.0056	0.3634	155.6463	0.0874	0.4794

Table 5. Results achieved with PNN trained by BBPSO

A comparison among the results obtained in this work with the found in (Oliveira et al., 2008) is done in Table 6. The results mentioned are the mean of the four subsets for each metric, and for those in bold are the best results found for each one of the metrics. It is important to highlight that such comparison is a little unfair, since the GA algorithm was executed with 80 individuals and 100 generations whereas the PSO and BBPSO were simulated with 50 particles and 100 iterations. Nevertheless, the results achieved to PNNs are similar. Furthermore, the approach using PSO and BBPSO got the best value of coverage and one-error, respectively. We can note that there is a discrepant difference among the performance of MLkNN and the performance obtained with the PNNs.

Again we can note a certain robustness of the PNN, because its performance didn't change significantly when trained by a PSO, BBPSO or GA algorithm.

Metrics	PNN-PSO	PNN-BBPSO	PNN-GA	ML-kNN-GA
Hamming loss	0.0055	0.0055	0.0055	0.0055
One-error	0.3752	**0.3604**	0.3736	0.4952
Coverage	**146.3293**	148.2525	156.4150	303.9029
Ranking loss	0.0889	0.0817	**0.0798**	0.1966
Average precision	0.4732	0.4863	**0.4880**	0.3813

Table 6. Comparison among different approaches of classification

7. Conclusions

The problem of classifying a large number of economic activities descriptions from free text format every day is a huge challenge for the Brazilian governmental administration. This problem is crucial for the long term planning in all three levels of the administration in Brazil. Therefore, an either automatic or semi-automatic manner of doing that is needed for making it possible and also for avoiding the problem of subjectivity introduced by the human classifier.

In this work, we presented an experimental evaluation of the performance of Probabilistic Neural Network on multi-label text classification. We performed a comparative study of probabilistic neural network trained by PSO and BBPSO, using a multi-label dataset for the categorization of free-text descriptions of economic activities. The approach using PSO and BBPSO were compared with GA and it was noted that there weren't significant differences among them.

To our knowledge, this is one of the first few initiatives on using probabilistic neural network for text categorization into a large number of classes as that used in this work and the results are very promising. One of the advantages of probabilistic neural network is that it needs only one parameter to be configured. In addition, the BBPSO employed is an almost parameter free algorithm, just the number of particles needs to be specified.

A direction for future work is to boldly compare the probabilistic neural network performance against other multi-label text categorization methods. Examining the correlation on assigning codes to a set of descriptions of economic activities may further improve the performance of the multi-label text categorization methods. We are planning on doing that in future work.

8. Acknowledgments

We would like to thank Andréa Pimenta Mesquita, CNAE classifications coordinators at Vitoria City Hall, for providing us with the dataset we used in this work. We would also like to thank Min-Ling Zhang for all the help with the ML-KNN categorization tool. Thank to all the colleagues, especially Alberto Ferreira De Souza, Claudine Badue, Felipe M. G. França and Priscila Machado Vieira Lima for their technical support and valuable comments on this work. This work is partially supported by the Internal Revenue Brazilian Service (Receita Federal do Brasil) and Fundação Espírito Santense de Tecnologia – FAPES-Brasil (grant 41936450/2008) for their support of this research work. Furthermore, R. A. Krohling thanks the partial funding of his research work provided by FAPES/MCT/CNPq (grant 37286374/2007).

9. References

Baeza-Yates, R. & Ribeiro-Neto, B. (1998). *Modern Information Retrieval*. Addison-Wesley, New York, 1st edition.

Clerc, M. & Kennedy, J. (2002). The Particle Swarm: Explosion Stability and Convergence in a Multi-dimensional Complex Space. *IEEE Transactions on Evolutionary Computation*, vol. 6:58 - 73.

CNAE (2003). Classificação Nacional de Atividades Econômicas Fiscal. *IBGE - Intituto Brasileiro de Geografia e Estatística*, Rio de Janeiro, RJ, 1.1 edition. http://www.ibge.gov.br/concla.

DNRC (2007). *Ranking das Juntas Comerciais Segundo Movimento de Constituição, Alteração e Extinção e Cancelamento de Empresas*. Ministério do Desenvolvimento, Indústria e Comércio Exterior - Secretaria do Desenvolvimento da Produção, Departamento Nacional de Registro do Comércio (DNRC).

Duda, R. O.; Hart, P. E. & Stork, D. G. (2001). *Pattern Classification*. Wiley-Interscience, New York, 2nd edition.

Dumais, S. T.; Platt, J.; Heckerman, D. & Sahami, M. (1998). Inductive Learning Algorithms and Representation for Text Categorization. In *Proceedings of the 7th ACM International Conference on Information and Knowledge Management*, pages 148–155, Bethesda, MD.

Eberhart, R. C. & Kennedy, J. (1995). A New Optimizer Using Particle Swarm Theory. *Proceedings of the Sixth International Symposium on Micro Machine and Human Science*, pages 39 - 43.

Fung, C. C.; Iyer, V.; Brown, W. & Wong, K. W. (2005). Comparing the Performance of Different Neural Networks Architectures for the Prediction of Mineral Prospectivity. *Proceedings of International Conference on Machine Learning and Cybernetics*, 1:pages 394 - 398.

Georgiou, V. L.; Pavlidis, N. G.; Parsopoulos, K. E.; Alevizos, P. D. & Vrahatis, M. N. (2004). Optimizing the Performance of Probabilistic Neural Networks in a Bioinformatic Task. *Proceedings of the EUNITE Conference*, pages pages 34 - 40.

Haykin, S. (1998). *Neural Networks - A Comprehensive Foundation*. Prentice Hall, New Jersey, 2nd edition.

Jain, A. K.; Murty, M. N. & Flynn, P. J. (1999). Data Clustering: a Review. *ACM Computing Surveys*, 31(3):264–323.

Kennedy, J. (2003). Bare Bones Particle Swarms. *Proceedings of the IEEE Swarm Intelligence Symposium*, pages 80 - 87.

Kennedy, J. & Eberhart, R. C. (1995). Particle Swarm Optimization. *Proceedings of the IEEE International Conference on Neural Networks* IV, pages 1942 - 1948.

Koster, C. H. A.; Seutter, M. & Beney, J. (2003). Multi-classification of Patent Applications with Winnow. *Ershov Memorial Conference*, vol. 2890, pages 546 - 555.

Krohling, R. A. & Coelho, L. S. (2006). Co-evolutionary Particle Swarm Using Gaussian Distribution to Solving Constraint Optimization Problems. *IEEE Transactions on Systems, Man and Cybernetics*, part B, vol. 36:1407 - 1416.

Li, X.; Chen, H.; Zhang, Z. & Li, J. (2007). Automatic Patent Classification using Citation Network Information: an Experimental Study in Nanotechnology. *Proceedings of the 2007 Conference on Digital Libraries*, pages 419 - 427.

Mao, K. Z.; Tan, K. C. & Ser, W. (2000). Probabilistic Neural-Network Structure Determination for Pattern Classification. *IEEE Transactions on Neural Networks*, 11:1009–1016.

Oliveira, E.; Ciarelli, P. M.; Souza, A. F. & Badue, C. (2008). Using a Probabilistic Neural Network for a Large Multi-label Problem. *10th Brazilian Symposium on Neural Networks*, pages 1 - 6.

Oliveira, E.; Ciarelli, P. M. & Lima, F. O. (2007). The Automation of the Classification of Economic Activities from Free Text Descriptions using an Array Architecture of Probabilistic Neural Network. *VIII Simpósio Brasileiro de Automação Inteligente*, pages 1 - 5.

Parzen, E. (1962). On Estimation of a Probability Density Function and Mode. *The Annals of Mathematical Statistics*, 33:pages 1065 – 1076.

Patra, P. K.; Nayak, M.; Nayak, S. K. & Gobbak, N. K. (2002). Probabilistic Neural Network for Pattern Classification. *IEEE Proceedings of the 2002 International Joint Conference on Neural Networks*, volume 2, pages 1200–1205.

Reynolds, C. W. (1987). Flocks, Herds, and Schools: A Distributed Behavioral Model. *Computer Graphics*, vol. 21(4):25 – 34.

Schapire, R. E. & Singer, Y. (2000). BoosTexter: A Boosting-based System for Text Categorization. *Machine Learning*, 39(2/3):135–168.

Sebastiani, F. (2002). Machine Learning in Automated Text Categorization. *ACM Computing* Surveys, 34(1):1–47.

Specht, D. F. (1990). Probabilistic Neural Networks. *Neural Networks*, 3(1):109–118.

Souza, A. F.; Pedroni, F.; Oliveira, E.; Ciarelli, P. M.; Henrique, W. F. & Veronese, L. (2007). Automated Free Text Classification of Economic Activities using VG-RAM Weightless Neural Networks. *7th International Conference on Intelligent Systems Design and Applications*, pages 1 - 5.

Zhang, M.-L. & Zhou, Z.-H. (2007). ML-KNN: A Lazy Learning Approach to Multi-Label Learning. *Pattern Recognition.*, 40:2038–2048.

11

A Novel Binary Coding Particle Swarm Optimization for Feeder Reconfiguration

Men-Shen Tsai and Wu-Chang Wu
National Taipei University of Technology
Taipei, Taiwan

1. Introduction

Power distribution systems are formed by many inter-connected feeders. Each feeder is further partitioned into many load-zones by switches. These switches can be divided into two categories: normally closed sectionalizing-switches and normally opened tie-switches. During normal operation, the structure of distribution system must be maintained in radial structure by properly adjusting the status of the switches. The distribution system can be reconfigured by changing the status of these switches while maintaining the radial structure. The feeder reconfiguration serves several purposes, for example, reducing power losses, maintaining load balance and enhancing service reliability. The mean of a switch operation plan is that by changing the status of sectionalizing-switches and tie-switches, loads can be transferred from one feeder to an adjacent feeder to redistribute loads without violating the operation limitations. However, great deals of switches exist on distribution systems. The number of possible solutions for feeder reconfiguration is increased in exponential order when the number of switches on distribution system increases. Thus selecting the best switch operation plan from all feasible solutions can be considered as an NP-Complete problem. Because the status of switches can be represented as '1' or '0', the problem of feeder reconfiguration can also be regarded as '1' and '0' permutation combinatorial optimization problems.

Researchers studied the feeder reconfiguration problems using different methods in the past decades. The results of these researches provide acceptable solutions for feeder reconfiguration problems. Heuristic methods to minimize power losses and improve the searching speed were proposed in (Baran & Wu, 1989). Soft computing approaches were applied to the problem extensively as well, for example, neural network (Kim et al., 1993), simulated annealing (SA) (Chang & Kuo, 1994), genetic algorithm (GA) (Nara et al., 1992; Kitayama & Matsumoto, 1995) and evolutionary programming (EP) (Hsiao, 2004; Hsu & Tsai, 2005). Algorithms based on concept of mimicking swarm intelligent are popular in recent years. For instance, ant colony optimization (ACO) (Teng & Lui, 2003; Carpaneto & Chicco, 2004; Khoa & Phan, 2006) and particle swarm optimization (PSO) (Chang & Lu, 2002) are the algorithms that can be applied to the field of optimization problems. These algorithms are applied to the problems of power distribution system gradually.

This research will apply the concept of PSO algorithm that is a novel and suitable algorithm for solving combinatorial optimization problems. Kennedy and Eberhart (Kennedy &

Eberhart, 1995; Shi & Eberhart, 1998) proposed PSO (typical PSO) in 1995. The PSO can be treated as the branch of the evolutionary algorithms and it introduces the concept of swarm intelligent. There are many similarities between PSO and Genetic Algorithm (GA). Both algorithms produce an initial solution set randomly at first. Through iterations of the evolution process, optimal solution can be obtained. The major difference between GA and PSO is that PSO has no explicit selection, crossover and mutation operations (Eberhart & Shi, 1998). Searching process in PSO is based on the previous best solution of a particle and the best solution of the population so far to update particle's information. That means the particles will share the best information between each other and lead the particles moving toward the target. Due to the searching mechanism designed in PSO, the probability of falling into local solution for PSO algorithm can be reduced. Also, the concept of PSO is simple and is easy to implement than GA. Thus, PSO can be a powerful algorithm to aid and speed up the decision-making process for feeder reconfiguration problems to identify the best switching plan.

As mentioned previously, feeder reconfiguration problems are non-linear discrete optimization problems. However, the typical PSO is designed for continuous function optimization problems; it is not designed for discrete function optimization problems. Fortunately, Kennedy and Eberhart proposed a modified version of PSO called Binary Particle Swarm Optimization (BPSO) that can be used to solve discrete function optimization problems (Eberhart & Kennedy, 1997). Although BPSO can be applied to solve the discrete optimization problems, there are still problems when BPSO is applied for feeder reconfiguration problems. In feeder reconfiguration problems, there are a large number of tie-switches. Randomly choosing the locations of these tie-switches will cause outages or non-radial structure in distribution systems. In (Chang & Lu, 2002), BPSO is used to solve the feeder reconfiguration problems and the method they proposed avoided the problem of unsuitable numbers of tie-switches. The concept of (Chang & Lu, 2002) is based on BPSO and the moving velocity of particle is defined in terms of probabilities. Instead of BPSO used in (Chang & Lu, 2002), this research tries to construct a more feasible discrete PSO scheme based on typical PSO for feeder reconfiguration. The method proposed in this research modifies the operators of PSO's formula based on the characteristics of both the status of switches and the shift operator to construct the binary coding particle swarm optimization for feeder reconfiguration. Minimizing total line losses and load balancing without violating operation constraints and maintaining radial structure are the two objective functions in this research. The simulations will be performed and the results are used to compare the proposed method, the method proposed in (Chang & Lu, 2002) and BPSO to verify the performance and effectiveness. A distribution system in Taiwan Power Company (TPC) is used in this study to verify the stability and usefulness of the proposed algorithm.

2. Problem Statement

There are all kinds of loads on distribution systems and these loads distributed non-evenly on the distribution feeders. The uneven load distribution on feeders may cause the conductor overloading or transformer load unbalancing on distribution systems during emergency operation. Fig. 1 is a simple 3-feeder distribution system. The ampacity of each feeder is 300A. The total loads on each feeder are 105A, 250A and 200A respectively. This configuration is considered as an unbalanced distribution system when the feeder loading is concerned. The feeder reconfiguration can be performed by opening/closing of sectionalizing-switches and tie-switches on distribution systems to reduce line losses or increase the system reliability. Therefore, feeder reconfiguration can redistribute the loads and is a common practice for the distribution system operators to avoid the problems of the conductor/transformer overloading or unbalancing on distribution feeders or transformers. Fig. 2 is the result of feeder reconfiguration from Fig. 1. The loads on each feeder are 185A, 190A and 180A respectively after reconfiguration. As a result, the system is operated in a more balanced way. However, some constraints should be considered during feeder reconfiguration. These constraints include: the radial structure of distribution system must be maintained, all zones must be served, feeder capacity should not be exceeded and feeder voltage profile should be maintained. As mentioned earlier, the feeder reconfiguration problems can be treated as '1' & '0' permutation combinatorial optimization problems. '1' represents a normally closed switch; while '0' represents a normally opened switch. Considering a simple system shown in Fig. 1, the order of switch permutation is sw1, sw2, ..., sw11 in turn. Thus, the status of switch permutation of the system in Fig. 1 can be expressed as [1 1 0 1 1 1 1 0 1 1 1]. The result of feeder reconfiguration is shown in Fig. 2, and the switch permutation becomes [1 1 1 0 1 1 1 1 0 1 1].

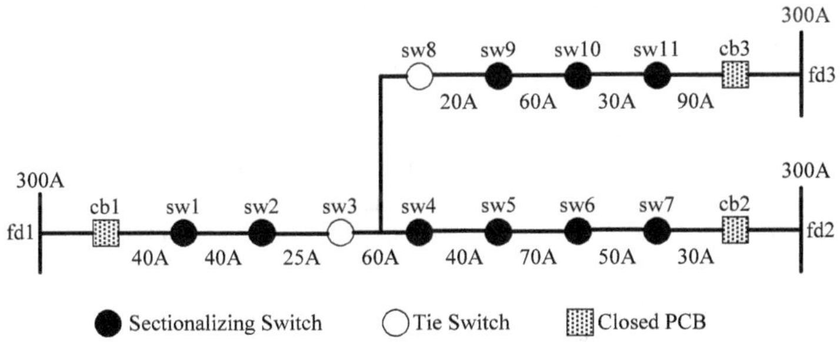

Figure 1. A simple 3-feeders distribution system

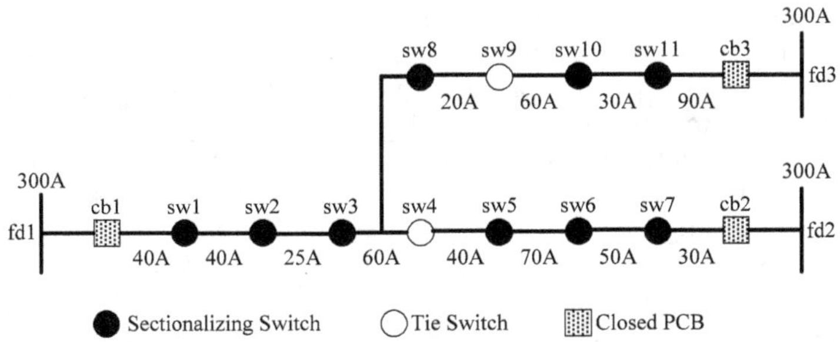

Figure 2. Result of feeder reconfiguration

Some objectives such as minimize the total line losses, minimize the numbers of operating switches, minimize voltage drop and load balance index are considered during feeder reconfiguration in general. Two objectives are considered in this research. The first is to minimize the total line losses during normal operation. By doing so, the operation of distribution system will be more economic and effective. The second objective is to distribute loads on feeders evenly. Balanced feeder loads can increase the opportunity of load transfer during emergency conditions and improve system reliability. The method proposed in this research also ensures that structure is maintained in radial and the ampacity of each conductor is kept within allowable limits. "Concentric load model" is used in this research for calculating branch currents. The line losses can be formulated as follows:

$$F_{loss} = \mathrm{Re}\left(\sum_{i=1}^{n} I_i^2 \cdot z_i\right) \tag{1}$$

where F_{loss} is the total real power losses of distribution feeders, n is the total numbers of zones in distribution system, I_i is the current magnitude of the i-th zone and z_i is the line impendence of the i-th zone. The load balance index is expressed as following:

$$F_{load\_balance} = \sum_{m=1}^{k}\sum_{n=1}^{k}(Cap_m - Cap_n)^2 \tag{2}$$

where k is number of feeder. Cap_m or Cap_n represents the total load of feeder m and n respectively. The total feeder loads can be calculated as following:

$$Cap_i = \sum_j Load_{i,j} \tag{3}$$

where, $Load_{i,j} \in Feeder_i$, i is the feeder number, and j is the load zone number within feeder i. In order to calculate the fitness value of the system represented by a particle, the method proposed in (Hsu & Tsai, 2005) is used to integrate the two object functions.

3. Particle Swarm Optimization

3.1 Typical Particle Swarm Optimization

A considerable amount of incredible social behavior and great intelligent exist in nature such as ant colonies, bird flocking, animal herding and fish schooling. Although the ability of individual is limited, the population can achieve the difficult target though cooperation with each other. Note that there is no centralized control in population. The behavior of individual depends on interacting with one another and with their environment only. These simple behaviors among individuals can lead population make themselves toward global behavior. Thus, completing a goal by aggregating the individuals and cooperating with each other that could be called swarm intelligent. Particle Swarm Optimization is one of the optimization algorithms provided with the concept of swarm intelligent. Original concept of PSO came from the study of simulating behavior of bird flocking to look for food. A possible solution for each problem can be represented as a particle that is just like a bird flocking in a D-dimensional searching space. Each individual particle has a fitness value that is evaluated by a fitness function to pick a good experience for itself and population respectively. The particles of population is initialized randomly first. A particle changed its searching direction based on two values or experiences during each iteration. The first one is the best searching experience of individual so far and it is called pbest. Another one is the best result obtained so far by any particle in the population and it is called gbest. When pbest and gbest are obtained, a particle updates its velocity and position based on (4) and (5). Lastly, the algorithm will check the results every iteration until the best solution is found or termination conditions are satisfied.

$$v_{id}^{new} = wv_{id} + c_1 \times rand() \times (pbest - x_{id}) + c_2 \times rand() \times (gbest - x_{id}) \quad (4)$$

$$x_{id}^{new} = x_{id} + v_{id}^{new} \quad (5)$$

In the above equations, v_{id} is the original velocity of the i-th particle, v_{id}^{new} is the new velocity of the i-th particle, w is the inertia weight, c_1 and c_2 are the acceleration constants, x_{id} is the original position of the i-th particle, x_{id}^{new} is the new position of the i-th particle and rand() is a random number ranging between 0 and 1.

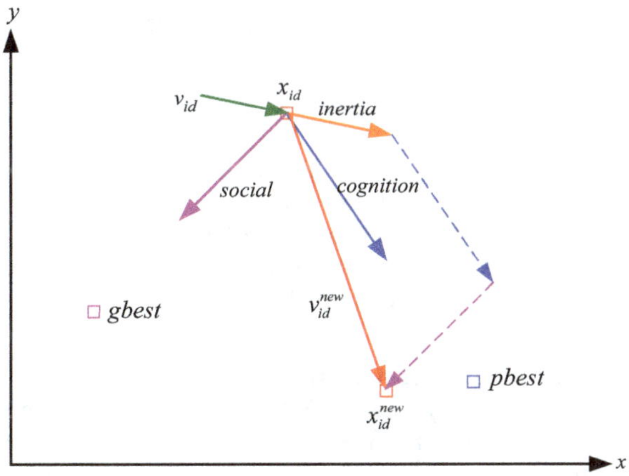

Figure 3. Searching diagram of typical PSO

In (4), the first part is the inertia (habitual behavior), which represents the particle trusts its own status at present location and provides a basic momentum. The second part is the cognition (self-knowledge) or memory, which represents the particle is attracted by its own previous best position and moving toward to it. The third part is the social (social knowledge) or cooperation, which represents the particle is attracted by the best position so far in population and moving toward to it. There are restrictions among these three parts and can be used to determine the major performance of the algorithm. The purpose of updating formula is to lead particles moving toward compound vector of inertia part, cognition part and social part. By doing so, the opportunity for particle to reach the target (optimal solution) will be increased. The inertia weight in the formula is used to adjust searching areas. A larger inertia weight will motivate the algorithm toward a global search; a smaller value will force the PSO toward a local search. The searching diagram of typical PSO is shown in Fig. 3.

3.2 Binary Particle Swarm Optimization

Kennedy and Eberhart proposed a binary version of PSO for discrete problems (Eberhart & Kennedy, 1997). In the binary PSO version, the particle's personal best and global best is still updated as in the typical version as described in (4). The elements inside x_{id}, pbest and gbest of BPSO are either `1' or `0'. Therefore, a particle flies in a search space restricted to zero and one. The speed of the particle must be constrained to the interval [0, 1]. A logistic sigmoid transformation function $S(v_{id}^{new})$ shown in (6) can be used to limit the speed of particle.

$$S(v_{id}^{new}) = \frac{1}{1+e^{-v_{id}^{new}}} \tag{6}$$

The update equation of BPSO can be done in two steps. First, (4) is used to update the velocity of the particle and the sigmoid function, (6), is used to limit the velocity in the interval [0, 1]. Second, the new position of the particle is obtained using (7) shown below:

$$\begin{aligned} if\left(rand() < S(v_{id}^{new})\right) \ then \ x_{id}^{new} &= 1 \\ else \ x_{id}^{new} &= 0 \end{aligned} \tag{7}$$

where, rand() is a uniform random number in the range [0, 1].
Since the relevant variables are derived from the changes of probabilities, the concept of BPSO is different from the typical PSO. It is hard to identify the relation between the current status and previous status of a particle. The selection of parameters, such as inertia weight, acceleration constants, etc., is also problematic.

3.3 Binary Coding Particle Swarm Optimization

Through the discussion of typical PSO and BPSO in the previous section, the PSO algorithm cannot be applied to feeder reconfiguration directly. Therefore, this research tries to construct a more feasible discrete PSO scheme based on the concept of typical PSO for feeder reconfiguration. The typical PSO must be modified based on the characteristics of distribution feeder operations. Two issues will be considered in the modification process. The first one is the problem of feeder reconfiguration is '1' & '0' permutation combinatorial optimization problem. The second issue is utilizing the shift operator that is used in computer programming languages. The shift operator and shift operator set defined in this research using these two aspects. Shift operator and shift operator set can be used to construct the binary coding particle swarm optimization for distribution feeder reconfiguration. These two definitions and the proposed binary coding PSO will be discussed.

3.3.1 Shift Operator

Suppose m sectionalizing switches (normally closed, N.C.) and n tie switches (normally opened, N.O.) exist on a distribution system. The permutation combination of the status all switches (s=m+n) is [S_1, S_2, …, S_s] and it will be called 'sequence of switch states', or SSS, in the rest of this paper. The shift operator is defined as SO (Bit_i, $Direction_{L,R}$, $Step_c$) and it means that an action will change the position of an N.O. in SSS. Bit_i is the index of i-th switch in SSS. $Direction_{L,R}$ indicates the direction of left or right shifting on the i-th switch. $Step_c$ is the number of shifting steps. The new permutation in SSS is defined as SSS'=SSS <+> SO. The symbol, '<+>', represents the shift operator. It will be applied to SSS to get a new SSS'.

A case is used to explain the operating process of shift operator. A simple distribution system shown in Fig. 4 has four feeders, nine N.C.s and three N.O.es. The SSS of this system is denoted as [1 0 1 0 1 1 1 1 1 1 0 1]. Supposing an SO(4, R, 1) is applied on this SSS. The process of operation is described as Fig. 5. When an N.O. shifts, a '1' (N.C.) needs to be set at its original position to maintain system structure.

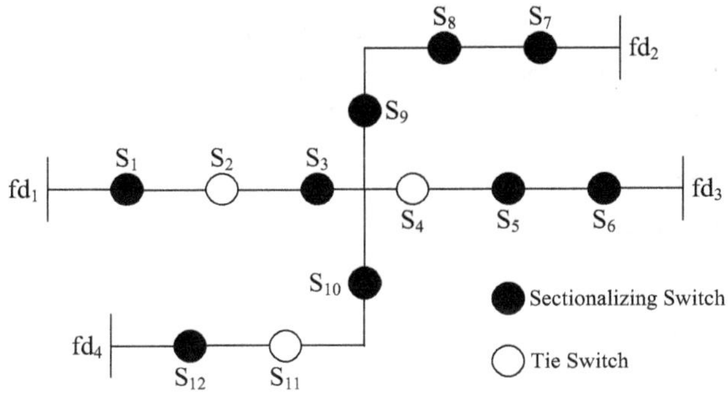

Figure 4. A simple 4-feeders distribution system

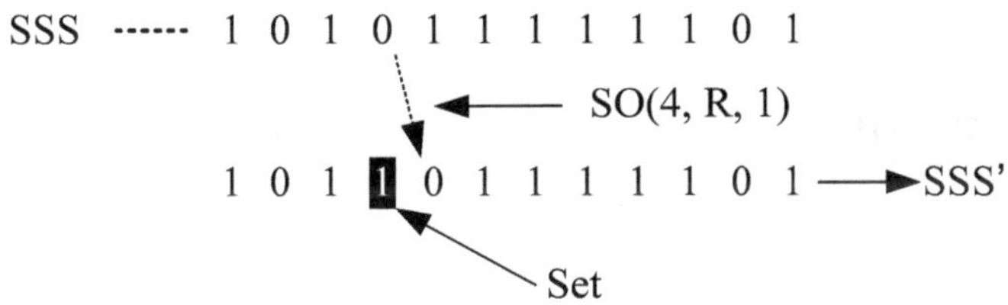

Figure 5. Basic operating process of shift operator

3.3.2 Shift Operator Set

A set with at least one or more shift operators is called shift operator set (SOS). An SOS represents all actions about how to set or shift normal open switches on distribution systems. The definition of shift operator set is shown in (8).

$$SOS = \{SO_1, SO_2, ..., SO_n\} \tag{8}$$

where n is the number of shift operators.

Considering two SSSes, SSS_1 and SSS_2, a set of shift operators which transfers SSS_1 to SSS_2 needs to be identified. Two SSSes, SSS_1=[1 0 1 0 1 1 1 1 1 1 0 1] and SSS_2=[1 1 1 1 0 1 1 0 1 0 1 1], are used to explain how the shift operators are obtained. By comparing the position of normally opened switch one by one in these two SSSes, the SOS can be acquired. The determination of the shift operator set and the result are shown as Fig. 6. In this example, SOS={ SO_1, SO_2, SO_3}= $SSS_2 \ominus SSS_1$. The symbol, '$\ominus$', is used to represent an action to get the shift operators from SSS_1 to SSS_2.

Base on the concept of above process, (pbest - x_{id}) and (gbest - x_{id}) in (4) can be rewritten as (pbest $\ominus$ x_{id}) and (gbest $\ominus$ x_{id}) respectively. The x_{id}, pbest and gbest represent different

SSSes in this sketch. This process will transfer an SSS to a new one which is closer to the best switch plan.

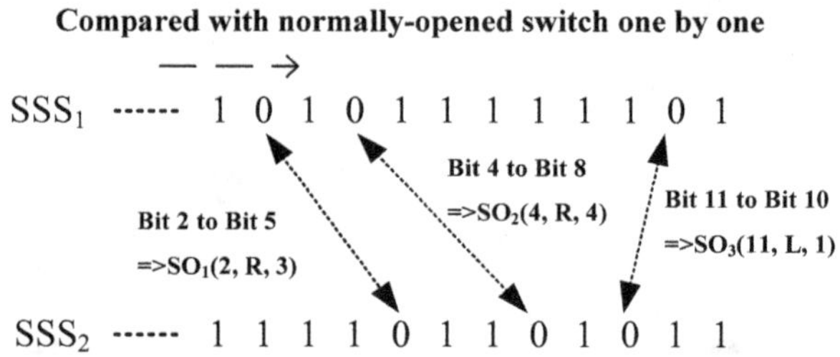

Figure 6. Decision process of shift operator set

3.3.3 Constructing Binary Coding PSO

The definition of shift operator and shift operator set are discussed in previous sections. The velocity update formulas (4) and (5) of PSO can be reestablished to solve the problem of feeder reconfiguration. The new velocity update formula for the proposed binary coding PSO is as below:

$$v_{id}^{new} = (w \otimes v_{id}) \oplus (rand() \langle \times \rangle (pbest \, \Theta \, x_{id})) \oplus (rand() \langle \times \rangle (gbest \, \Theta \, x_{id})) \tag{9}$$

$$x_{id}^{new} = x_{id} \langle + \rangle v_{id}^{new} \tag{10}$$

The symbol, '$\oplus$', shown in (9) is used for combining two shift operator sets. The symbol, '$\otimes$', is the operator that is used to shift the number of steps. The symbol, '$\langle \times \rangle$', is used to select the number of shift operator, SO, in (pbest Θ x_{id}) or (gbest Θ x_{id}) randomly. x_{id} is the original SSS of the i-th particle; pbest is the best SSS of the i-th particle; gbest is the best SSS of any particle in the population. v_{id} is the original shift operator set of the i-th particle, v_{id}^{new} is the new shift operator set of the i-th particle. x_{id}^{new} is the new SSS of the i-th particle. rand() is a random number with a range of [1, n] where n is the number of SO in SOS.

In Eq. (9), w is the inertia weight. The role of w is used for adjusting searching areas. The searching areas are reduced progressively when the number of iteration increases. The inertia weight can be calculated as (11).

$$w = \frac{iteration_{\max} - iteration_{now}}{iteration_{\max}} \times ShiftStep_{\max} \tag{11}$$

A simple example is used to show how the proposed method works. Based on the system shown in Fig. 4, x_{id}, pbest and gbest represent different SSSes are given:

x_{id} : [1 0 1 0 1 1 1 1 1 1 0 1]
pbest : [1 1 1 1 0 1 1 0 1 0 1 1]
gbest : [1 1 0 1 0 1 1 1 0 1 1 1]

The SOS can be derived from (pbest Θ x_{id}) and (gbest Θ x_{id}) as:

$$(\text{pbest} \,\Theta\, x_{id}) = \{(2, R, 3), (4, R, 4), (11, L, 1)\}$$
$$(\text{gbest} \,\Theta\, x_{id}) = \{(2, R, 1), (4, R, 1), (11, L, 2)\}$$

The three parts in (9) can be expressed as following:

$$w \otimes v_{id} = \{(2, L, 3), (4, L, 2), (11, R, 2)\}$$
$$\text{rand}() \langle\times\rangle (\text{pbest} \,\Theta\, x_{id}) = \{(2, R, 3), (4, R, 4), (11, L, 1)\}$$
$$\text{rand}() \langle\times\rangle (\text{gbest} \,\Theta\, x_{id}) = \{(2, R, 1), (11, L, 2)\}$$

According to (9), the v_{id}^{new} contains eight SOes, (2, L, 3), (2, R, 3), (2, R, 1), (4, L, 2), (4, R, 4), (11, R, 2), (11, L, 1) and (11, L, 2). Combining these eight SOes, the final v_{id}^{new} contains three SOes, (2, R, 1), (4, R, 2) and (11, L, 1). Finally the new SSS, x_{id}^{new}, will be [1 1 0 1 1 0 1 1 1 0 1 1] according to (10).

The procedure of proposed binary coding PSO is outlined as below:

a. Determine the size of population and other parameters such as number of iterations and maximum shift steps.
b. Initialize the SSS and shift operator sets randomly to produce particles.
c. Evaluate the fitness value for each particle.
d. Compare the present fitness value of i-th particle with its historical best fitness value. If the present value is better than pbest, update the information including SSS and fitness value of pbest.
e. Compare present fitness value with the best historical fitness value of any particle in population. If the present fitness value is better than gbest, update the information including SSS and fitness value for gbest.
f. Update the shift operator set and generate a new SSS of the particle according to (9) and (10), respectively.
g. If stop criterion is satisfied then stop, otherwise go to step c). In this research, the stop criterion is the iteration count reaches the maximum number of iteration.

4. Experimental Results

To verify the performance of the proposed algorithm and compare with algorithms of typical BPSO (Eberhart & Kennedy, 1997) and modified BPSO (Chang & Lu, 2002) for feeder reconfiguration problem, a four-feeder distribution system is used. This distribution system is taken from Taoyuan division, Taiwan Power Company, Taiwan. The system has 24 sectionalizing-switches, 8 tie-switches and 28 load-zones, as shown in Fig. 7. The capacity of each feeder is shown in Table 1. The objective functions are: minimizing feeder loss and load balancing index without violating operation constraints. The proposed method and the algorithms described in (Eberhart & Kennedy, 1997) and (Chang & Lu, 2002) were implemented using Java language for comparison purposes. Relevant parameters are set as follows. The size of population is 10 for all methods. Maximum number of iteration is set to 1000 for all methods as well. The inertia weight, learning factor of c_1 and c_2 for the methods of typical BPSO (Eberhart & Kennedy, 1997) and modified BPSO (Chang & Lu, 2002) are set to 0.8, 2.0 and 2.0, respectively. The settings of these parameters can be referred to (Chang & Lu, 2002). In order to obtain the results and calculate the average performance, 10 runs were performed for each method.

The comparisons of the results from the three algorithms are shown in Table 2. The Max, Min and Average in Table 2 indicate the maximum, minimum and average fitness value, running time, losses and load balancing index values in 10 runs respectively. The typical BPSO is not able to get a better result than proposed algorithm due to the higher probability of inadequate number of tie-switches represented by particles. Although the running time of typical BPSO is less than proposed method, the average values of losses and load balancing index of typical BPSO are higher than proposed method. The modified BPSO is able to avoid the problem of inadequate number of tie-switches represented in each particle. On the other hand, the result of proposed method is better than other two methods. Beside the execution time of proposed method is two seconds longer than BPSO, all the other outcomes of proposed method are superior to other methods. The feeders which represent of maximum fitness value of feeder reconfiguration of the typical BPSO method, modified BPSO and proposed method are shown in Fig. 8, Fig. 9 and Fig. 10 respectively. Table 3 lists the comparison of total loads of each feeder obtained from the three methods. All the results indicate that the proposed method provides better and more reliable solutions than typical BPSO and modified BPSO methods for minimizing line losses and load balancing problem.

Feeder ID	F1	F2	F3	F4
Capacity (Amp)	500	500	250	500

Table 1. Capacity of each feeder

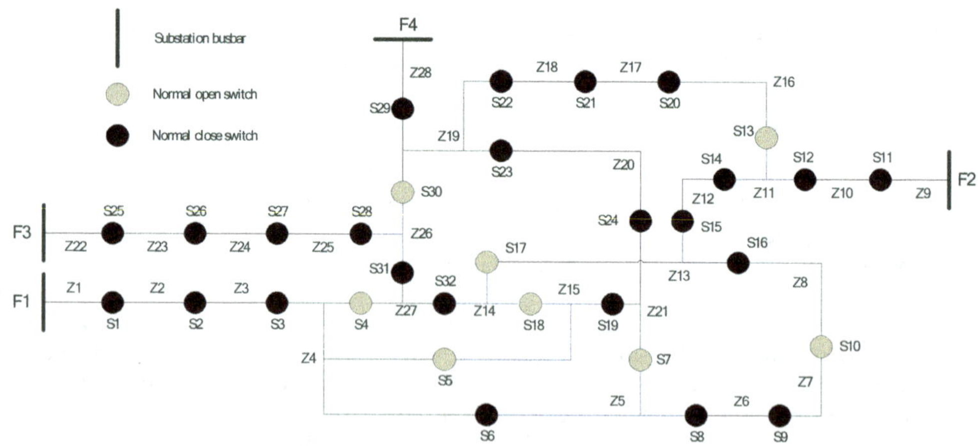

Figure 7. A 4-feeders distribution system for testing

Method		Typical BPSO	Modified BPSO	Binary coding PSO
Fitness Value	Max	0.8759	0.9121	0.9234
	Min	0.8058	0.8844	0.8898
	Average	0.8594	0.8992	0.9032
Running Time (msec)	Max	6625	11015	8734
	Min	5250	8812	8110
	Average	6212	10359	8354
Loss	Max	515kW	405kW	364kW
	Min	339kW	335kW	312kW
	Average	404kW	365kW	329kW
Load Balance Index	Max	525928	434216	264648
	Min	184712	183368	169112
	Average	329504	294859	208328

Table 2. Results and comparisons of three algorithms

Method \ Feeder ID	F1	F2	F3	F4
Original system	176	146	171	203
Typical BPSO	124	312	122	138
Modified BPSO	139	232	122	203
Binary Coding PSO	139	227	110	220

Table 3. The comparison of the feeder loading

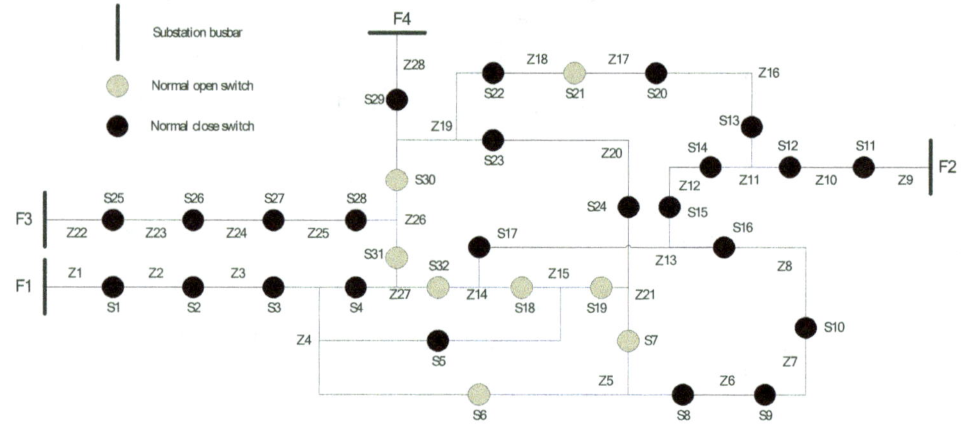

Figure 8. The final feeder configuration found by the typical BPSO method

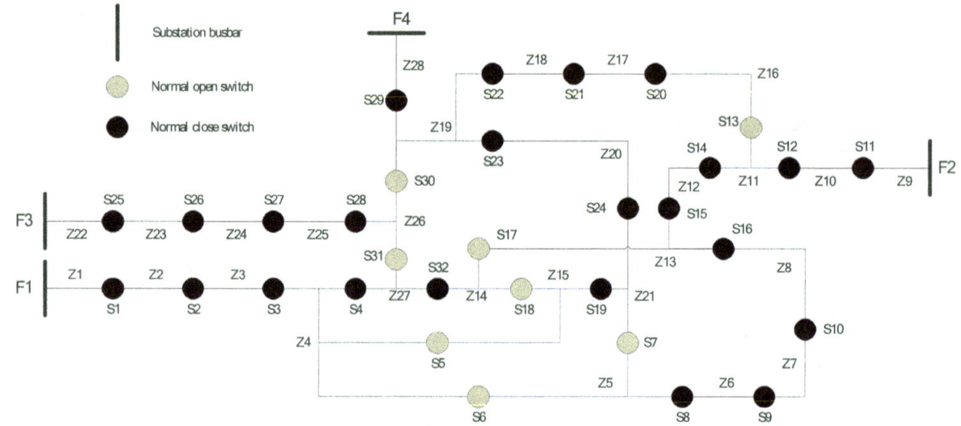

Figure 9. The final feeder configuration found by the modified BPSO method

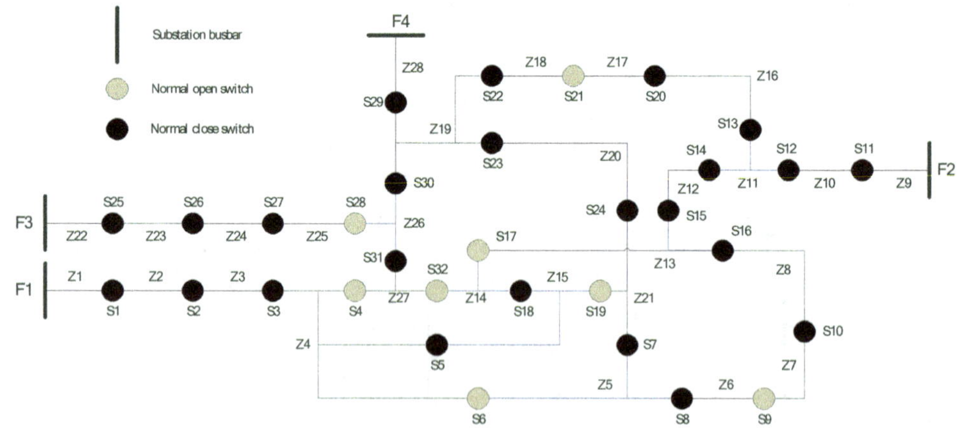

Figure 10. The final feeder configuration found by the proposed method

5. Conclusion

Particle Swarm Optimization is a novel and powerful algorithm for continuous and discrete functions optimization problems. In this work, the concept of typical PSO is modified and applied to the feeder reconfiguration problems. Feeder reconfiguration problems are non-linear discrete optimization problems in nature; and further, there are some defects to use typical BPSO directly for feeder reconfiguration. This research try to construct a binary coding particle swarm optimization based on typical PSO to solve this problem. The operators of typical PSO algorithm have been reviewed and redefined in this research to fit the application of distribution feeder reconfiguration. In addition, minimizing total line losses and load balancing without violating operation constraints are the objective functions used in this research. The experimental results show that the proposed method can solve the feeder reconfiguration problem more effectively.

6. Acknowledgement

This research was supported by the National Science Council, Taiwan under contract NSC 97-2221-E-027-110.

7. References

Baran M.E. and Wu F.F. (1989). Network Reconfiguration in Distribution Systems for Loss Reduction and Load Balancing, *IEEE Trans. on Power Delivery,* vol. 4, no.2, April 1989, pp. 1401-1407.

Chang H. C. and Kuo C. C. (1994). Network reconfiguration in distribution system using simulated annealing, *Elect. Power Syst. Res,* vol. 29, May 1994, pp. 227-238.

Chang R.F. and Lu C.N. (2002). Feeder Reconfiguration for Load Factor Improvement, *IEEE Power Engineering Society Winter Meeting,* Vol. 2, 27-31 Jan. 2002, pp.980-984.

Carpaneto E. and Chicco G. (2004). Ant-Colony Search-Based Minimum Losses Reconfiguration of Distribution Systems, *Proc. IEEE Melecon 2004,* pp.971-974, Dubrovnik, Croatia.

Eberhart R.C. and Kennedy J. (1997). A Discrete Binary Version of the Particle Swarm Algorithm, *In Proceedings of IEEE International Conference on Systems, Man, and Cybernetics,* vol. 5, pp.4104-4108, 1997.

Eberhart R.C. and Shi Y. (1998). Comparison between Genetic Algorithms and Particle Swarm Optimization, *The 7th Annual Conference on Evolutionary Programming,* San Diego, USA, 1998.

Hsiao Ying Tung. (2004). Mutiobjective Evolution Programming Method for Feeder Reconfiguration, *IEEE Trans. on Power Systems,* Vol. 19, No. 1 pp. 594-599, February 2004.

Hsu Fu-Yuan and Tsai Men-Shen. (2005). A Multi-Objective Evolution Programming Method for Feeder Reconfiguration of Power Distribution System, *Proc. of the 13th Conf. on Intelligent Systems Application to Power Systems,* pp.55-60, November 2005.

Kim H., Ko Y. and Jung K.H. (1993). Artificial Neural Networks Based Feeder Reconfiguration for Loss Reduction in Distribution Systems, *IEEE Trans. on Power Delivery,* vol. 8, no. 3, July 1993, pp. 1356-1366.

Kennedy J. and Eberhart R. C. (1995). Particle Swarm Optimization, *Proceedings IEEE Int'l. Conf. on Neural Networks, IV*, pp.1942-1948, 1995.

Kitayama M. and Matsumoto K. (1995). An Optimization Method for Distribution System Configuration Based on Genetic Algorithm, *Proceedings of IEE APSCOM*, pp. 614-619, 1995.

Khoa T.Q.D. and Phan B.T.T. (2006). Ant Colony Search based loss minimum for reconfiguration of distribution systems, *2006 IEEE Power India Conference.* Page(s): 6pp, April 2006.

Nara K., Shiose A., Kitagawa M. and Ishihara T. (1992). Implementation of Genetic Algorithm for Distribution Systems Loss Minimum Reconfiguration, *IEEE Trans. on Power Systems*, vol.7, no. 3, August 1992, pp. 1044-1051.

Shi Y. and Eberhart R.C. (1998). A modified particle swarm optimizer, *IEEE International Conference on Evolutionary Programming*, pp.69-73, May 1998, Alaska.

Teng Jen-Hao and Lui Yi-Hwa (2003). A Novel ACS-Based Optimum Switch Relocation Method, *IEEE Trans. on Power Systems*, vol. 18, no. 1, February 2003, pp.113-120.

12

A Particle Swarm Optimisation Approach to Graph Permutations

Omar Ilaya and Cees Bil
RMIT University
Australia

1. Introduction

In many real-world applications, the arrangement, ordering, and selection of a discrete set of objects from a finite set, is used to satisfy a desired objective. The problem of finding optimal configurations from a discrete set of objects is known as the *combinatorial optimisation problem*. Examples of combinatorial optimisation problems in real-world scenarios include network design for optimal performance, fleet management, transportation and logistics, production-planning, inventory, airline-crew scheduling, and facility location.

While many of these combinatorial optimisation problems can be solved in polynomial time, a majority belong to the class of NP -hard (Aardal et al., 1997). To deal with these hard combinatorial optimisation problems, approximation and heuristic algorithms have been employed as a compromise between solution quality and computational time (Festa and Resende, 2008). This makes heuristic algorithms well-suited for applications where computational resources are limited. These include dynamic ad-hoc networks, decentralised multi-agent systems, and multi-vehicle formations. The success of these heuristic algorithms depends on the computational complexity of the algorithm and their ability to converge to the optimal solution (Festa and Resende, 2008). In most cases, the solutions obtained by these heuristic algorithms are not guaranteed optimal.

A recently developed class of heuristic algorithms, known as the *meta-heuristic algorithms*, have demonstrated promising results in the field of combinatorial optimisation. Meta-heuristic algorithms represent the class of all-purpose search techniques that can be applied to a variety of optimisation problems including combinatorial optimisation. The class of meta-heuristic algorithms include (but not restricted to) simulated annealing (SA), tabu search, evolutionary algorithms (EA) (including genetic algorithms), ant colony optimisation (ACO) (Aguilar, 2001), bacterial foraging (Passino, 2002), scatter search, and iterated local search.

Recently, a new family of computationally efficient meta-heuristic algorithms better posed at handling non-linear constraints and non-convex solution spaces have been developed. From this family of meta-heuristic algorithms, is particle swarm optimisation (PSO) (Kennedy and Eberhart, 1995). Like other biologically inspired meta-heuristic algorithms, PSO is an adaptive search technique that is based on the social foraging of insects and animals. In PSO, a population of candidate solutions are modelled as a swarm of particles. At each iteration, the particles update their position (and solution) by moving stochastically

towards regions previously visited by the individual particle and the collective swarm. The simplicity, robustness, and adaptability of PSO, has found application in a wide-range of optimisation problems over continuous search spaces. While PSO has proven to be successful on a variety of continuous functions, limited success has been demonstrated to adapt PSO to more complex richer spaces such as combinatorial optimisation.
In this chapter, the concepts of the standard PSO model are extended to the discrete combinatorial space and a new PSO is developed to solve the combinatorial optimisation problem. The chapter is organised as follows: In Section 2, a brief review of related works to solving the combinatorial optimisation space using meta-heuristics is presented. In Section 3, the standard PSO model is introduced. The nature of the combinatorial optimisation problem is then presented in Section 4 before the concepts of the standard PSO model are adapted to the combinatorial space in Section 5. Section 6 analyses the stability and performance of the newly developed algorithm. The performance of the newly developed algorithm is then compared to the performance of a traditional genetic algorithm in Section 7 before Section 8 concludes with final remarks.

2. Related Works

In recent years, variants of traditional PSO have been used to solve discrete and combinatorial optimisation problems. A binary PSO was first developed in (Kennedy and Eberhart, 1997) to solve discrete optimisation problems. In the binary PSO, each particle encoded a binary string in the solution space. A particle moved according to a probability distribution function determined using the Hamming distance between two points in the binary space. The early concepts introduced by the binary PSO appeared in later PSO algorithms for combinatorial optimisation such as in (Shi et al., 2006); (Tasgetiren et al., 2004); (Liu et al., 2007b); (Pang et al., 2004); (Martínez García and Moreno Pérez, 2008); (Song et al., 2008); and (Wang et al., 2003). Tasgetiren et al. (Tasgetiren et al., 2004) introduced the smallest position value rule (SPV) to enable the continuous PSO algorithm to be applied the class of sequencing and combinatorial problems. In SPV, each particle assigns a position value in continuous space to each dimension in the discrete space. At each iteration, the position value is updated according to the traditional velocity update equation and the sequence of objects is re-sorted according to the values assigned to the continuous space. The method proposed by (Tasgetiren et al., 2004) is similar to the random keys in GA (Bean, 1994). Following a similar method to (Kennedy and Eberhart, 1997), Wang et al. (Wang et al., 2003) introduced the concept of a swap operator to exchange dimensions in the particle position. In (Wang et al., 2003), each particle encoded a permutation of objects and a transition from one position to the next was achieved by exchanging elements in the permutation. To account for both the personal best positions and global best positions, Wang et al. extended the concept of swap operator to swap sequence. The swap sequence was used to move a particle from one position to the next by successively applying a sequence of swap operators. Using this approach, the notion of velocity on the combinatorial space was defined; and the Hamming distance was used to exclusively determine the motion of a particle. Premature convergence was addressed by randomly applying the swap operator to the particle. Similar approaches to Wang et al. include (Shi et al., 2006); (Martínez García and Moreno Pérez, 2008); and (Bonyadi et al., 2007), where a swap sequence was also constructed through the concatenation of successive swap operators. The ordering of these swap operators influences the position of the particle at the end of each iteration. In (Wang et al., 2003); (Shi et al., 2006); (Martínez García and Moreno Pérez, 2008); and (Bonyadi et al., 2007), the swap sequence is constructed by first applying the swap operators that move the particle to it's personal best, followed by the swap operators that move the particle to it's global best. For sufficiently small perturbations, the particles will tend towards the global best position of the swarm and stimulate the loss of solution diversity. This invariably leads to the rapid convergence of the algorithm and poor solution quality. For large complex optimisation problems, the PSO must compromise the local and global search strategies effectively to find high-quality (if not optimal) solutions rapidly. In addition, the PSO framework must be sufficiently robust to adapt to a wide variety of discrete and combinatorial optimisation problems. In this chapter, a generalised combinatorial optimisation framework is introduced that builds on the works of (Wang et al., 2003); (Shi et al., 2006); (Tasgetiren et al., 2004); and (Kennedy and Eberhart, 1997) to develop a new combinatorial optimisation PSO. In the following section, a brief introduction into the traditional PSO is presented before the main results of this chapter are developed.

3. The Standard Particle Swarm Optimisation Model

Let $\mathcal{P}$ denote a D-dimensional problem, and $f(x): X \to \mathbb{R}$ an objective function for the problem that maps X to the set of real numbers. Without loss of generality, consider the following optimisation problem $x^* \in X \Leftrightarrow x^* = \arg\min_{x \in X} f(x) \quad \forall x \in X$. In traditional PSO, a solution i is represented by a particle in a swarm P moving through D-dimensional space with position vector $x_k^i = (x_k^i(1), \dots, x_k^i(d), \dots, x_k^i(D))$ for any time k. At each iteration, the particles adjust their velocity v_k^i along each dimension according to the previous best position of the i-th particle p_k^i and the best position of the collective swarm p_k^g (see Fig. 1). The position x_k^i for the i-th particle is updated according to the following velocity function:

$$v_{k+1}^i = w \cdot v_k^i + c_1 \cdot r_1 \cdot (p_k^i - x_k^i) + c_2 \cdot r_2 \cdot (p_k^g - x_k^i) \tag{1a}$$

$$x_{k+1}^i = x_k^i + v_k^i \tag{1b}$$

where $r_1, r_2 \in [0,1]$ are random variables affecting the search direction, $c_1, c_2 \in \mathbb{R}$ are configuration parameters weighting the relative confidence in the personal best solutions and the global best solutions respectively, and w is an inertia term influencing the momentum along a given search direction. Algorithm 1 summarises the iterative nature of the PSO algorithm.

The terms c_1 and c_2 are the main configuration parameters of the PSO that directly influence the convergence of the algorithm. For large values of c_1, exploration of particles is bounded to local regions of the best previously found solutions p_k^i. This maintains population diversity and is favourable when the problem is characterised by non-linear and non-convex solution spaces. In contrast, large c_2 values will encourage particles to explore regions closer to the global best solution p_k^g at each iteration. Generally, this search strategy will converge faster and is practical for convex solution spaces with unique optima. Adjusting the inertia term w affects the relative weighting of the local and global searches. A large w encourages the particles to explore a larger region of the solution space at each iteration and maximise global search ability, whilst a smaller w will restrict the particles to local search at each iteration (Shi and Eberhart, 1998b).

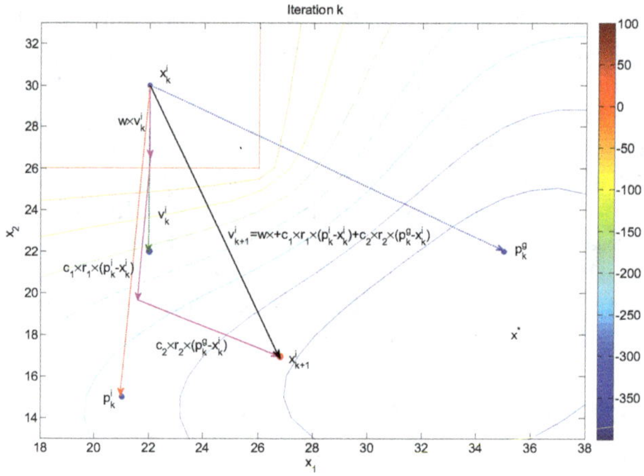

Figure 1. Particle position and velocity on a two-dimensional vector space

---

for all particle i **do**
    initialise position x_k^i randomly in the search space
end for
while termination criteria not satisfied **do**
    for all particle i **do**
        set personal best p_k^i as the best position found by the particle so far
        set global best p_k^g as the best position found by the swarm so far
    end for
    for all particle i **do**
        update velocity according to
        $v_{k+1}^i = w \cdot v_k^i + c_1 \cdot r_1 \cdot (p_k^i - x_k^i) + c_2 \cdot r_2 \cdot (p_k^g - x_k^i)$
        update position according to
        $x_{k+1}^i = x_k^i + v_k^i$
    end for
end while

Algorithm 1. Traditional PSO

4. Problem Description and Model Construction

The combinatorial optimisation problem for PSO is now discussed. Let $X = \{x^1, x^2, \ldots, x^i, \ldots\}$ denote the finite set of solutions to the combinatorial optimisation problem with objective function $f : X \rightarrow \mathbb{R}$. Assume the objective of the combinatorial optimisation problem is to find $x^* \in X$, such that $x^* \in X \Leftrightarrow x^* = \arg\min_{x \in X} f(x) \quad \forall x \in X$. Consider the case where a solution $x^i \in X$ to the combinatorial optimisation problem is given by the linear ordering of elements in the set $[n]=\{1,2,\ldots,n\}$, such that $\forall x^i \in X$, $x^i=(x^i(1),x^i(2),\ldots,x^i(n)) \in \{1,2,\ldots,n\}$. Then $|X|=n!$. Each integer value in the list encodes the relative ordering of a set of objects and is referred to as a *permutation* of objects (Bóna, 2004). These include cities in a tour, nodes in a network, jobs in a schedule, or vehicles in a formation. For convenience, a permutation is represented using *two-line form*. Let $g:[d]\rightarrow[n]$ be a bijection on the ordered list. If $[n]$ describes the list of numbers $[n]=\{1,\ldots,n\}$, then $[d]=\{1,\ldots,n\}$ and g is also a permutation of the set $[n]$ (Bóna, 2004).

Example 1.

As an example, consider the following permutation $\{3,4,1,5,2\}$. The function $g:[5]\rightarrow[5]$ defined by $g(1)=3$, $g(2)=4$, $g(3)=1$, $g(4)=5$, and $g(5)=2$ is also permutation of $[5]$ (Bóna, 2004). In two-line form, the set $[5]$ can be written as:

$$g = \begin{matrix} 1 & 2 & 3 & 4 & 5 \\ 3 & 4 & 1 & 5 & 2 \end{matrix}$$

where it is implied that g maps 1 to 3, 2 to 4, 3 to 1, 4 to 5, and 5 to 2.

5. Fitness Landscape

In order to adapt PSO to the combinatorial space, it is convenient to define a metric space characteristic of the combinatorial optimisation problem. Let $\mathcal{N}: X \rightarrow 2^X$ denote a syntactic neighbourhood function that attaches to each solution $x^i \in X$ the neighbouring set of solutions $x^j \in \mathcal{N}_i(x^i) \subseteq X$ that can be reached by applying a unitary syntactic operation moving $x^i \mapsto x^j$ (Moraglio and Poli, 2004). Denote this unitary syntactic operator by φ and assume that the operation is reversible, i.e. $x^j \in \mathcal{N}_i(x^i) \Leftrightarrow x^i \in \mathcal{N}_j(x^j)$. Such a neighbourhood can be associated to an undirected *neighbourhood graph* $G=(V,E)$, where V is the set of vertices representing the solutions $x^i \in X$, and E the set of edges representing the transformation paths for permutations. By definition, the combinatorial space endowed with a neighbourhood structure $\mathcal{N}_i(x^i)$ and induced by a distance function $h_{ij}(x^i,x^j)$ is a *metric space*. Formally, the definition of a metric or distance function is any real valued function $h_{ij}(x^i,x^j)$ that conforms to the axioms of identity, symmetry, and triangular inequality, i.e.:

1. $h_{ij}(x^i,x^j)\geq 0$ and $h_{ij}(x^i,x^i)=0$ (identity);
2. $h_{ij}(x^i,x^j)=h_{ij}(x^j,x^i)$ (symmetric);
3. $h_{ij}(x^l,x^j)\leq h_{li}(x^l,x^i)+h_{ij}(x^i,x^j)$ (triangle inequality);
4. if $i\neq j$, then $h_{ij}(x^i,x^j)>0$.

A neighbourhood structure $\mathcal{N}_i(x^i)$ induced by a distance function $h_{ij}(x^i,x^j)$ can then be formally expressed as:

$$\mathcal{N}_i(x^i)=\{x^j \mid x^j \in X, h_{ij}(x^i,x^j)\leq s\} \tag{2}$$

where $s \in \mathbb{R}$. On a combinatorial space with syntactic operator φ, any configuration x^i can be transformed into any other x^j by applying the operator φ a finite number of times ($1 < s \leq n$) (Misevicius et al., 2004). In such a case, the distance metric $h_{ij}(x^i, x^j)$ is given by the Hamming distance:

$$h_{ij}(x^i, x^j) = \sum_{l=1}^{n} \text{sgn}\left|x^i(d) - x^j(d)\right|$$

and s represents the minimum number of exchanges to transform x^i into x^j. Other distance metrics can be similarly defined (see (Ronald, 1997); (Ronald, 1998); and (Moraglio and Poli, 2004) references therein for a comprehensive treatment on distance metrics defined on the combinatorial space).

For generality, only the *deviation distance metric* (Ronald, 1998) will be considered hereafter. While other distance metrics can be defined for discrete and combinatorial spaces, the decision to use the deviation distance metric is trivial with respect to algorithmic design. Other problem-specific metrics can be substituted into the developed algorithm with little influence on the procedural implementations of the algorithm.

The deviation distance metric provides a measure of the relative distance of neighbouring elements between two permutations x^i and x^j. In problems where the adjacency of two elements influences the cost of the objective function $f(x)$, such as in TSP and flow-shop scheduling, the deviation distance function provides an appropriate choice of metric for the problem space (Ronald, 1998). Formally, the positional perturbation Δ_a of one element value $x^i(d_1)$ to its matching value in $x^j(d_2)$, such that $x^i(d_1) = x^j(d_2) = a$, $a \in [n]$, is given by the following:

$$\Delta_a = |d_1 - d_2| \tag{3}$$

For convenience, Δ_a is normalised $\overline{\Delta}_a \in [0,1]$:

$$\overline{\Delta}_a = \frac{\Delta_a}{n-1} \tag{4}$$

The deviation distance $h_{ij}(x^i, x^j)$ is then defined as the sum of the $\overline{\Delta}_a$ values:

$$h_{ij}(x^i, x^j) = \sum_{a}^{n} \overline{\Delta}_a \tag{5}$$

From Eq. (5) a large position deviation induces a greater distance in the metric space. The notion of position deviation is now used to construct the combinatorial optimisation PSO.

6. Proposed Algorithm

In Section 4.1, the concept of a syntactic operator φ was discussed as a method of transforming one configuration x^i to another $x^j \in \mathcal{N}_i(x^i)$. In the following section, the parallel between a syntactic operator φ and the motion of a particle i in the combinatorial space is described. Let $x^i \in X$ encode a permutation of $[d]$ objects in D-dimensional space. The position $x^i \in X$ of a particle i in the D-dimensional space corresponds to a permutation of $[d]$ objects. Define φ by a two-way perturbation (transformation) operator $\varphi := SO(d_1, d_2)$ as the *swap operator* that exchanges elements d_1 and d_2 in solution x^i, such that $X \to X$, $d_1, d_2 \in \{1,2,\ldots D\}$, $d_1 \neq d_2$. Applying the swap operator to the permutation x^i, the following solution is derived:

$$x_{k+1}^i = x_k^i \oplus SO(d_1, d_2) \tag{6}$$

where $x^i(d_1) = x^j(d_2) = a$, and $x_k^j, x_k^{i+1}, x^j \in \mathcal{N}_i(x_k^i)$, and the notation $\oplus$ is used to denote x_{k+1}^i is obtained from x_k^i by applying the perturbation $SO(d_1, d_2)$. In the combinatorial optimisation PSO, x_k^i and $x_k^j \in \mathcal{N}_i(x_k^i) \subseteq X$, $x_k^i \neq x_k^j$ encode two permutations in the combinatorial optimisation problem and represents positions in the combinatorial search space. Applying the notions of swap operator to PSO, the swap operator $SO(d_1, d_2)$ for a particle i can be interpreted as a motion of the particle x_k^i to a position x_k^j displaced from x_k^i by the deviation distance $h_{ij}(x_k^i, x_k^j)$. Consider the case when $x_k^j \notin \mathcal{N}_i(x_k^i)$. Then, the following transition $x_k^i \mapsto x_k^j$ is not possible by Eq. (6) alone. Define the following *swap sequence* (Knuth, 1998):

$$SS = \{SO_1, SO_2, \ldots, SO_n\} \tag{7}$$

where SS is the concatenation of swap operators and the order of the swap operators SO_i, $i = 1,\ldots,n$ is influential to the final position x_{k+1}^i. The minimum number of swap operators required to move $x_k^i \mapsto x_k^j$ is given by the Hamming distance and is referred to as the *basic swap sequence* (Knuth, 1998).

Suppose particle i moves according to $x_k^i \mapsto p_k^i$. The basic swap sequence transforming x_k^i to p_k^i can be determined by moving along each dimension of the initial position x_k^i and applying the Partially Mapped Crossover function (PMX) (Goldberg and Lingle, 1985) to each dimension along x_k^i. The PMX function maps each dimension in the current position x_k^i to the corresponding dimension in p_k^i (see Fig. 2). A swap operator is invoked if the object in the d_1-th dimension of the p_k^i solution and the x_k^i are inconsistent. The d_1-th element in x_k^i is then swapped with the d_2-th element in x_k^i such that $x_k^i(d_2) = p_k^i(d_1)$. Algorithm 2 summarises the basic swap operator used to move $x_k^i \mapsto p_k^i$

while $d(x^i, x^j) \neq 0$
    if $x_k^i(d_1) \neq x_k^j(d_1)$ **then**
        find d_2 such that $x_k^i(d_2) = x_k^j(d_1) = a_1$, and $d_1, d_2 \in \{1,\ldots,D\}$
            set $SO_j(d_1, d_2)$ and store as j-th entry in SS
    else, end if
end while

Algorithm 2. Basic Swap Operator

Note, applying the algorithm from left-right gives $d_2 > d_1$, $d_1, d_2 \in \{1,2,\ldots,D\}$.

Example 2.

Consider the following two solutions $x^i = \binom{1\ 2\ 3\ 4\ 5}{1\ 2\ 3\ 4\ 5}$ and $x^j = \binom{1\ 2\ 3\ 4\ 5}{2\ 3\ 1\ 5\ 4}$ represented in two-line form. Applying Algorithm 2 from left to right, the first swap operator is invoked if $x^i(1) \neq x^j(1)$. Since $x^i(1) = 1$ and $x^j(1) = 2$, the following mapping is observed between object $1 \rightarrow 2$. The first swap operator is then given by the exchange of elements 1 and 2 in x^i, $SO_1(1,2)$. Following $SO_1(1,2)$, particle i is now at position $x' = \binom{1\ 2\ 3\ 4\ 5}{2\ 1\ 3\ 4\ 5}$. Comparing x' to x^j, the following mapping $1 \leftrightarrow 3$ is now observed between object $x'(2)$ and $x^j(2)$. The next mapping is then given by $SO_2(2,3)$ taking x' to $x'' = \binom{1\ 2\ 3\ 4\ 5}{2\ 3\ 1\ 4\ 5}$. Repeating this procedure, the swap sequence SS that takes x^i to x^j is then given by $SS = \{SO_1(1,2), SO_2(2,3), SO_3(4,5)\}$ such that $x^j = x^i \oplus SS$.

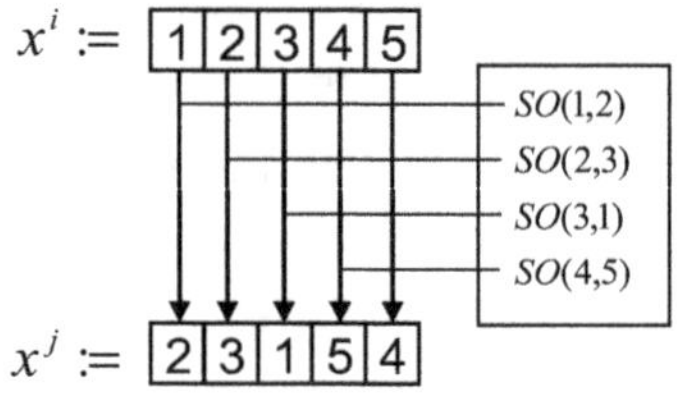

Figure 2. Partially-mapped crossover (PMX)

In traditional PSO, the motion of a particle is influenced by the personal best position p_k^i and global best of the swarm p_k^g. In the combinatorial optimisation PSO, each position encodes a permutation to the combinatorial optimisation problem. If the personal best and global best positions are not coincident, i.e. $p_k^i \neq p_k^g$, then the swap sequences SS_1 and SS_2 that moves the i-th particle along the transformations $x_k^i \mapsto p_k^i$ and $x_k^i \mapsto p_k^g$ respectively, are not equivalent, i.e. $SS_1 \neq SS_2$. Application of SS_1 or SS_2 will yield $x_{k+1}^i = p_k^i$ or $x_{k+1}^i = p_k^g$ and will cause the particles to converge towards the personal best solution, or the global best solution respectively. This leads to rapid convergence and sub-optimal solution quality. The local search induced by the exclusive application of SS_1, and the global search induced by the exclusive application of SS_2 is now combined to develop a velocity update function with similar characteristics to the original PSO algorithm.

In the traditional PSO algorithm, the velocity of a particle is composed of three parts; the *momentum term*, i.e. $v \cdot w$, the *cognitive velocity* $c_1 \cdot r_1 \cdot (p_k^i - x_k^i)$, and the *social velocity* $c_2 \cdot r_2 \cdot (p_k^g - x_k^i)$. Using the notions of *momentum, cognitive velocity,* and *social velocity,* the following decoupled velocity update for a particle in the combinatorial space with deviation distance metric Δ_a is defined:

$$v_{k+1}^{l,i} = w \cdot v_k^{l,i} + c_1 \cdot (\Delta_a'(x_k^i, p_k^i)) \tag{8a}$$

$$v_{k+1}^{g,i} = w \cdot v_k^{g,i} + c_2 \cdot (\Delta_a'(x_k^i, p_k^g)) \tag{8b}$$

where w, c_1, and c_2 have the same meanings as the original PSO algorithm. For convenience, denote Eq. (8a) as the *local velocity* and Eq. (8b) as the *global velocity*. Equation (8a) and (8b) preserve the same tuning parameters as the original PSO without the random variables $r_1, r_2 \in [0,1]$. The decision to omit the random variables is trivial, but will become apparent in the proceeding section.

Recall, the position of each particle x_k^i, $\forall i \in P$ is a vector in the D-dimensional combinatorial space $x_k^i \in X$ and moves along the dimensions of the D-dimensional hypercube by exchanging elements via the swap operator $SO(d_1, d_2)$. The velocity of each particle v_k^i, $\forall i \in P$ is a vector in the D-dimensional *continuous* space $v_k^i \in \mathbb{R}^D$ and describes the local gradient of the fitness landscape using the deviation distance metric. Using the velocity $v_k^i \in \mathbb{R}^D$, a probability mapping is described that invokes the swap operator and preserves the contributions of both the local velocity and global velocity. Let $\Pr(x^i(d) \mid p^i(d))$ and $\Pr(x^i(d) \mid p^g(d))$ denote the sampling probability of the i-th particle for dimension d in the particle when the individual best is $p^i(d)$ and global best is $p^g(d)$ respectively. Then, the probability that $x^i(d)$ moves to $p^i(d)$ and $p^g(d)$ is given by the following statements:

$$\Pr(x_k^i(d) \mid p_k^i(d)) := v_k^{l,i} \tag{9a}$$

$$\Pr(x_k^i(d) \mid p_k^g(d)) := v_k^{g,i} \tag{9b}$$

Since $p^i(d)$ and $p^g(d)$ is a mapping for $x^i(d) \mapsto p^i(d)$ and $x^i(d) \mapsto p^g(d)$ respectively, the probability that the swap operator $SO_j(d_1, d_2)$ is invoked by moving $x^i(d) \mapsto p^i(d)$ or $x^i(d) \mapsto p^g(d)$ using Algorithm 2 is defined using the local and global velocities respectively:

$$\Pr(SO(d_1, d_2)) := v_k^{l,i} \tag{10a}$$

$$\Pr(SO(d_1, d_2)) := v_k^{g,i} \tag{10b}$$

where $x^i(d_2) = p^i(d_1)$ or $x^i(d_2) = p^g(d_1)$ for $x^i(d) \mapsto p^i(d)$ and $x^i(d) \mapsto p^g(d)$ respectively. Following Eq. (10a) and Eq. (10b), the velocity $v_k^i(d)$ describes the probability that an element in $x_k^i(d)$ will swap with the corresponding element in $x_k^j(d)$ and invoke Algorithm 2, then the velocity on each dimension $d \in D$ must be bounded over the interval $v_k^i(d) \in [0,1]$. The velocities described in Eq. (10a) and Eq. (10b) are normalised according to:

$$v_{k+1}^{l,i} = \frac{v_{k+1}^{l,i}}{\arg\max\{v_{k+1}^{l,i}, v_{k+1}^{g,i}\}} \tag{11a}$$

$$v_{k+1}^{g,i} = \frac{v_{k+1}^{g,i}}{\arg\max\{v_{k+1}^{l,i}, v_{k+1}^{g,i}\}} \tag{11b}$$

Normalising the velocities with respect to both the personal best and global best velocity profiles is used to prioritise the order of swap operations and preserve the probability map. Once an element $x_k^i(d_1)$ has been swapped with the corresponding element $x_k^i(d_2)$ in $p_k^i(d_1)$, the associated velocity $v_k^i(d_2)$ at element $x_k^i(d_2)$ is set to zero if $x_k^i(d_2) = p_k^i(d_2)$ to prevent cyclic behaviour.

Using the definition of the sample probability in Eq. (10a) and Eq. (10b) for the personal best and global best respectively, the swap sequence induced by the combinatorial optimisation PSO can now be described. From Eq. (8a) and Eq. (8b), large deviation distances incur a large velocity. This observation is complimentary to the original concepts of the traditional PSO algorithm. Following Eq. (10a) and Eq. (10b) a large velocity will induce a greater probability that a swap operation is invoked with either the personal best or global best. Using this concept, a swap sequence can be defined using the relative probabilities of the personal best and global best velocity profiles. Consider the case when $v_k^{l,i}(d) > v_k^{g,i}(d)$. Then, the probability of exchanging $x_k^i(d) \mapsto p_k^i(d)$ is greater than the probability of exchanging $x_k^i(d) \mapsto p_k^g(d)$. In the swap sequence, the larger of the two probabilities will receive a higher priority in the swap sequence and take precedence over the lower probability swap operations. At a given iteration, particle i will move according to the following swap sequence:

$$x_{k+1}^i = x_k^i \oplus SS \tag{12}$$

where $SS = (SO_1(x_k^i(d), p_k^i(d)), SO_2(x_k^i(d), p_k^g(d)))$ if $v_k^{l,i} > v_k^{g,i}$. Algorithm 3 describes the implementation of the swap sequence SS.

---

for all $d \in D$ **do**
  if $v_k^{l,i}(d) > v_k^{g,i}(d)$ **do**
    invoke swap operator $SO_j(d_1, d_2)$ for $x_k^i \mapsto p_k^i$ using Algorithm 2
      if $x_k^i(d_2) = p_k^i(d_2)$ **do**
        set $v_k^{l,i}(d_2) = 0$
      else, end if
    goto 8
  otherwise if $v_k^{g,i}(d) > v_k^{l,i}(d)$ **do**
    invoke swap operator $SO_j(d_1, d_2)$ for $x_k^i \mapsto p_k^g$ using Algorithm 2
      if $x_k^i(d_2) = p_k^g(d_2)$ **do**
        set $v_k^g(d_2) = 0$
      else, end if
    goto 2
  end if
end for

Algorithm 3. Swap Sequence

Following the definition of the basic swap sequence and swap sequence in Algorithm 2 and Algorithm 3 respectively, the proposed combinatorial PSO algorithm can now be defined. Algorithm 4 describes the procedural implementation of the swap sequence within the context of the traditional PSO algorithm.

7. Algorithmic Analysis

The behaviour of each particle in the swarm can be viewed as a traditional line-search procedure dependent on a stochastic step size and a stochastic search direction. Both the stochastic step size and search direction depend on the selection of social and cognitive parameters. In addition, the stochastic search direction is driven by the best design space locations found by each particle and by the swarm as a whole. Unlike traditional line-search procedures however, PSO uses information from neighbouring particles to influence the search direction at each iteration. This exchange of information plays an important role in the stability and performance of the swarm. In the following section, the spectral properties of algebraic graph theory are used to show that for a fully interconnected swarm, the particles will reach a consensus on the equilibrium. The analysis begins by considering the original PSO algorithm with velocity and position update given by Eq. (1a) and Eq. (1b).

---

for all particle i **do**
  initialise position x_k^i randomly in the search space
end for
while termination criteria not satisfied **do**
  for all particle i **do**
    set personal best p_k^i as the best position found by the particle so far
    set global best p_k^g as the best position found by the swarm so far
  end for
  for all particle i **do**
    update local velocity according to
    $v_{k+1}^{l,i} = w \cdot v_k^{l,i} + c_1 \cdot (\Delta_v'(x_k^i, p_k^i))$
    update global velocity according to
    $v_{k+1}^{g,i} = w \cdot v_k^{g,i} + c_2 \cdot (\Delta_v'(x_k^i, p_k^g))$
    normalise local velocity according to
    $v_{k+1}^{l,i} = v_k^{l,i} / \arg\max\{v_k^{l,i}, v_k^{g,i}\}$
    normalise global velocity according to
    $v_{k+1}^{g,i} = v_k^{g,i} / \arg\max\{v_k^{l,i}, v_k^{g,i}\}$
    update position according to
    $x_{k+1}^i = x_k^i \oplus SS$
    where SS is determined from Algorithm 3
  end for
end while

Algorithm 4. Combinatorial Optimisation PSO

Without loss of generality, consider the following objective function for the combinatorial optimisation problem:

$$x^* = \arg\min_{x \in X} f(x) \ \forall x \in X$$

Then, the personal best p_k^i is the current best solution of the i-th particle found so far; i.e. $p_k^i = \arg\min_\tau x_\tau^i$, $\forall \tau \in (0,k]$; and the global best p_k^g is the current best solution of the global swarm found so far; i.e. $p_k^g = \arg\min_\tau x_\tau^i$, $\forall \tau \in (0,k]$, $\forall i \in N$. The swarm of particles is said to have reached an equilibria if and only if all the particles have reached a consensus on the value of p_k^g, i.e., $p_k^l = p_k^g = p^e$. For asymptotic convergence, all the particles in the swarm must globally asymptotically reach a consensus on the global best solution, such that $x^e = \lim_{k \to +\infty} x_k^i$, and $x_k^{e,i} = x_k^{e,j} = \min(X), \forall i,j \in X, i \neq j$. For convenience, Eq. (1a) and Eq. (1b) are combined into compact matrix form:

$$\begin{bmatrix} x_{k+1}^i \\ v_{k+1}^i \end{bmatrix} = \begin{bmatrix} -(c_1 r_1 + c_2 r_2) & w \\ -(c_1 r_1 + c_2 r_2) & w \end{bmatrix} \cdot \begin{bmatrix} x_k^i \\ v_k^i \end{bmatrix} + \begin{bmatrix} c_1 r_1 & c_2 r_2 \\ c_1 r_1 & c_2 r_2 \end{bmatrix} \cdot \begin{bmatrix} p_k^i \\ p_k^g \end{bmatrix} \tag{13}$$

which can be considered as a discrete-dynamic system representation of the original PSO algorithm.

7.1 Equilibrium of the PSO

Before the main analysis results are presented, a brief introduction into algebraic graph modelling of swarms is presented. The information flow in the swarm of particles can be represented using an interconnected graph $G = (V,E)$, where V is the enumerated set of particles $x_k^i \in V$, $i \in \{1,\ldots,N\}$ in the swarm, and $E \subseteq V \times V$ is the set of edge relations between neighbouring particles. The *order* $|V|$ and *size* $|E|$ of the graph G physically represents the number of particles in the swarm and the number of edge connections. For a fully connected swarm, each particle communicates with every other particle in the population, and the graph is said to be complete. This is the case of the original PSO algorithm. The connectivity of a graph is described by the square matrix A, with size $|V|$, and elements a_{ij} describing the connectivity of adjacent vertices x^i and x^j, such that:

$$a_{ij} = \begin{cases} 1, & \text{if } (x^i, x^j) \in E \\ 0, & \text{otherwise} \end{cases} \tag{14}$$

The matrix A uniquely defines the connectivity of the graph G and is referred to as the *adjacency matrix*. Associated with the adjacency matrix A is the *graph Laplacian* L, and its *Laplacian potential* Ψ_G:

$$L = \Lambda^{-1}(\Lambda - A) \tag{15}$$

$$\Psi_G = \frac{1}{2} x^T L x \tag{16}$$

where Λ is the square matrix containing the out-degree of each vertex along the diagonal, and x is the concatenation of particles in the swarm. A well-known property of the Laplacian potential is that it is positive semi-definite and satisfies the following sum-of-squares property (Godsil and Royle, 2001):

$$x^T L x = \sum_{i,j \in E} A_{ij} \left(x^j - x^i \right)^2, \quad x \in \mathrm{R}^n \tag{17}$$

Using Eq. (17), the objective is to show that the personal best positions of each particle reaches a consensus (by way of equilibria) coincident to the global best of the swarm, i.e. $p_k^i = p_k^g$, $\forall \in P$. Eq. (16) becomes:

$$p^T L p = \sum_{i,j \in E} A_{ij} \left(p_k^j - p_k^i \right)^2, \quad p_k^i \in \mathrm{R}^n \tag{18}$$

where p is the concatenated states of the personal best of each of the particles in the swarm. The closed-loop dynamics of the global best position evolve according to the following continuous-time dynamic equation:

$$\dot{p} = -Lp = -\nabla \Psi_G \tag{19}$$

The equilibrium points of Eq. (19) correspond to stationary points of Ψ_G and the region outside of these points, the potential is strictly decreasing (Moreau, 2004); i.e., if x^e is an equilibrium of Eq. (18), then $Lx^e = 0$. From Eq. (16):

$$\Psi_G(p^e) = \frac{1}{2}(p^e)^T L p^e = 0 \tag{20}$$

Following the connectivity of G, $p_i^e = p_j^e = c, \forall i, j \in N$, i.e. $p^e = (c,\ldots,c)^T, c \in X$. Since the Laplacian potential equals zero at equilibrium, then $p^g = \min(p)$ is an invariant quantity, Given the invariance property of $\min(p)$, then $\min(p^e) = \min(p(0))$, and $\min(p^e) = c$. This implies $p_k^{e,i} = \min(p(0))$, $\forall i \in P$ (Olfati-Saber and Murray, 2003). This leads to the following observations for the particle dynamics in Eq. (1a) and Eq. (1b) that are consistent with the works of (Clerc and Kennedy, 2002); (Trelea, 2003); and (Kadirkamanathan et al., 2006):

1. The system dynamics are stochastic and order two;
2. The system does not have an equilibrium point if $p_k^g \neq p_k^l$;
3. If $p_k^l = p_k^g = p^e$ is time invariant, there is a unique equilibrium at $v^e = 0$, $x^e = p^e$.

An equilibrium point thus exists only for the best particle whose local best solution is the same as the global best solution (Kadirkamanathan et al., 2006).
Consider the case for a given particle i when the external input is constant (as is the case when no personal or global better positions are found). From Eq. (15) the eigenvalues of $-L$

are negative in the complex plane. Then, for particle i, the position asymptotically converges to the point x^e in the eigenspace associated to the global minimum found by the swarm of particles (Olfati-Saber and Murray, 2003). Such a position x^e is not necessarily a local or global minimiser of the combinatorial optimisation problem. Instead, it will improve towards the optimum x^* if a better individual or global position is found. Discovery of better individual or global positions can be improved by increasing the population diversity of the swarm through the introduction of chaos or turbulence (Kennedy and Eberhart, 1995). In the following section, a non-stationary Markov chain is constructed to integrate the discrete syntactic swap operators introduced in Section 5 to the continuous time-dynamics of the traditional PSO

7.2 Non-Stationary Markov Model of combinatorial PSO

Markov chains are important in the theoretical analysis of evolutionary algorithms operating on discrete search spaces (Poli et al., 2007) and have been used to model the probabilistic convergence of population-based meta-heuristic algorithms (see (Rudolph, 1996); (Cao and Wu, 1997); (Poli and Langdon, 2007); and (Greenwood and Zhu, 2001) for examples of their implementation). While traditional PSO has operated on a continuous search space, the combinatorial PSO operates on a discrete combinatorial space. This makes Markov chains a suitable method of modelling and analysing the behaviour of the combinatorial PSO. The use of Markov chains on bare-bones PSO has previously been investigated by (Poli and Langdon, 2007) where the continuous search space was discretised using a hypercube sampling. In the following section, a non-stationary Markov chain is used to model the combinatorial PSO and account for the newly introduced swap operator.

Let X denote the finite state space describing the set of permutation encodings with $r=|X|=n!$ possible solutions. Let $P \subset X$ be a population of solutions from X with size $|P|=N$. Then a finite Markov chain $\Gamma \subseteq X$ describes a probabilistic trajectory over the finite state space X (Rudolph, 1996) with $\mathbf{N}=\binom{m+n!-1}{n!-1}$ possible populations as states; i.e.:

$$X=\{S_1,S_2,\ldots,S_{\mathbf{N}}\} \tag{21}$$

The probability $q^{mn}_{k-1,k} := \Pr_{mn}(\Gamma^k = S_n \mid \Gamma^{k-1} = S_m)$ of transitioning from state $S_m \in X$ to $S_n \in X$, $m,n \in \mathbf{N}$ at step k is called the *transition probability* from m to n at step k. The transition probability of a finite Markov chain can be gathered into a *transition matrix* $Q_k = \{q^{mn}_{k-1,k}\}$ (Rudolph, 1994), where each dimension $q^{mn}_{k-1,k} \in [0,1]$. In a *stationary Markov chain* the probabilities remain fixed, and the Markov chain is said to be *homogenous;* i.e., $Q=Q_k=\{q^{mn}_{k-1,k}\}$, $\forall k=1,2,\ldots$, and $m,n=1,2,\ldots,\mathbf{N}$. In the case of the combinatorial PSO, the probabilities of the swap operator are updated according to Eq. (8a) and Eq. (8b). This results in a *non-stationary Markov chain.* The transition probabilities of non-stationary Markov chains are calculated by considering how the population incidence vector S_j describes the composition of the next iteration (Cao and Wu, 1997). Denote $z^{l,i}_k = \Pr(x^i_k \mid p^i_k) = v^{l,i}_k$ as the sampling probability when the personal best is p^i_k; likewise, denote $z_k^{g,i} = \Pr(x_k^i | p_k^g) = v_k^g$ as the sampling probability when the global best is p_k^g. The probability that a particle i will move according to $x_k^i \mapsto p_k^i$ or $x_k^i \mapsto p_k^g$ x_k^i is given by $z_k^i = (p_k^i \cup p_k^g)$. From Algorithm 3, the dimension for $z_k^{l,i} = \Pr(x_k^i | p_k^i)$ and $z_k^{g,i} = \Pr(x_k^i | p_k^g)$ is calculated independently using Eq. (8a) and Eq. (8b) and the probability of a particle sampling p_k^i or p_k^g is given by:

$$\Pr(p_k^i \cup p_k^g) = \prod_{d=1}^{D} \Pr\left(p_k^i(d) \cup p_k^g(d) \middle| p_k^i(d), p_k^g(d)\right) \tag{22a}$$

$$\Pr(p_k^i \cup p_k^g) = \prod_{d=1}^{D} \Pr(p_k^i(d)) + \Pr(p_k^g(d)) - \Pr(p_k^i(d)) \cdot \Pr(p_k^g(d)) \tag{22b}$$

Since personal bests can only change if there is a fitness improvement, only certain state transitions can occur. That is, a transition from state $S_m \mapsto S_n$ is possible only if the fitness of at least one particle in the swarm improves (Poli and Langdon, 2007). Because of the independence of the particles (over one time step), the state transition probability for the whole PSO is given by:

$$q_{k-1,k}^{mn} = \prod_i \Pr(p_k^i \cup p_k^g) \tag{23}$$

From Sec. 6.1, the local velocity $v_k^{l,i}$ and global velocity $v_k^{g,i}$ will tend to zero as $k \to +\infty$. This implies $\lim_{k\to\infty} q_{k-1,k}^{mn} = \lim_{k\to\infty} v_k^{l,i} = \lim_{k\to\infty} v_k^{g,i} = 0$, $m,n = 1,2,\ldots,\mathbf{N}$. Therefore, the swap operator preserves the convergent behaviour of traditional PSO and the combinatorial PSO converges to the equilibrium pair (x^e, v^e).

8. Numerical Examples

8.1 The Travelling Salesman Problem

To test the efficiency of the proposed algorithm, the combinatorial optimisation PSO is tested on the travelling salesman problem (TSP). TSP is an invaluable test problem that belongs to the class of NP-hard combinatorial optimisation problems. The objective of TSP is to find a minimum-cost tour that visits a set of n cities and returns to an initial point (Applegate et al., 2006). Mathematically, TSP is a combinatorial optimisation problem on an undirected graph $G = (V, E)$. Each city $c_i \in [n]$, $i = 1,2,\ldots,n$, is represented by a vertex $v_i \in V$ in the graph $G = (V, E)$ with cost of travel between adjacent cities given by $h_{ij} \in E$. A solution to TSP can be represented as a sequence of cities encoded by a permutation $x \in X$. Mathematically, the objective of TSP is given by the following optimisation problem:

$$x^* \in X \Leftrightarrow x^* = \arg\min_{x\in X} \sum_{d=1}^{n-1} h_{ij}(x(d), x(d+1)) + h_{ij}(x(n), x(1)) \tag{24}$$

Various problems, including path-finding, routing, and scheduling, can be modelled as a TSP. A repository of test-instances (and their solutions) is available through the TSPLIB library (Reinelt, 1991). In the following section, the combinatorial PSO is tested on several instances of the TPSLIB library. Table 1 summarises the test instances of TSPLIB used to validate and compare the combinatorial PSO.

Name	Dimension	Optimal $f(x)$	Optimal Solution
burma	14	30.8785	
gr17	17	2085	
gr24	24	1272	
eil51	51	426	

Table 1. Test instances taken from TSPLIB (Reinelt, 1991) used for the validation of the combinatorial PSO

8.2 Optimisation Results and Discussion

In the following experiments, the combinatorial PSO is applied to each case of the TSP in Table 1. The parameters used in each experiment are selected based on the findings reported in the literature (Zhang et al., 2005); (Shi and Eberhart, 1998a); (Zheng et al., 2003); (Clerc and Kennedy, 2002); and (Eberhart and Shi, 2000). While the inertia weight, cognitive and social parameters are sensitive to the problem domain in traditional PSO, a parametric analysis of their influence on the combinatorial PSO is beyond the scope of this chapter. For illustrative purposes, the parameters given in Table 2 are considered throughout the remainder of this chapter. The influence of these parameters on the performance of the combinatorial PSO remains the subject of future research.

Parameter	Value
w	0.8
c_1	2.025
c_2	2.025

Table 2. Combinatorial PSO parameters

To demonstrate the relative efficiency of the proposed algorithm, the performance of the combinatorial PSO is compared to a genetic algorithm. Each TSP experiment was trialled 100 times using randomly generated individuals. In both algorithms, a population of $P = 30$ was maintained for each iteration. The fitness values obtained by the combinatorial PSO and the GA over the 100 trials are presented in Table 3. Table 4 compares the success rate of the PSO and GA for each of the problems. Figure 3 compares the percentage of the solution space explored by the combinatorial PSO and the GA. This is determined as the number of unique solutions tested $x_k^i \in X$, $\forall i \in P$, $k = 1,\ldots,1000$ by the PSO and GA versus the size of the solution space $|X| = n!$.

Problem	Optimal Solution	Minimum		Maximum		Average	
		PSO	GA	PSO	GA	PSO	GA
burma	30.87	30.87	30.87	30.87	34.62	30.87	31.20
gr17	2085	2085	2085	2687	2489	2141.55	2175.02
gr24	1272	1272	1282	1632	1810	1453.52	1488.68
eil51	426	494.80	495.46	687.52	671.85	573.55	573.95

Table 3. Performance of the proposed algorithm compared to a traditional genetic algorithm for combinatorial optimisation

From Table 3, the combinatorial PSO outperformed the GA in all problem instances, except for the 51 variable eil51 problem. In this case, both the GA and combinatorial PSO failed to find the best solution over the 100 trials. Examination of Fig. 3 suggests that both the combinatorial PSO and GA were only able to search a small percentage ($\ll 1\%$) of the total solution space over the 1000 iterations. This suggests, that both the combinatorial PSO and GA experience a loss of solution diversity over the optimisation procedure. Figure 3 also indicates that the GA was able to cover a larger percentage of the solution space for each trial than the combinatorial PSO. This suggests that the combinatorial PSO suffers from the same rapid convergence and stagnation issues of traditional PSO. Loss of solution diversity and rapid convergence is a well-known problem in traditional PSO. In traditional PSO, the performance of the algorithm deteriorates as the number of iterations increases. Once the algorithm has slowed down (becomes stagnant), it is usually difficult to achieve a better fitness value; particularly for high-dimensionality problem spaces.

Problem	Success Rate (%)	
	PSO	GA
burma	100	92
gr17	36	17
gr24	4	0
eil51	0	0

Table 4. Success rate of the combinatorial PSO and GA

Recently, several methods have been proposed to improve solution diversity and avoid stagnation in traditional PSO. These methods include the use of chaos variables (Fieldsend and Singh, 2002); (Kennedy and Eberhart, 1995); and (He et al., 2004); variable neighbourhood topologies (Kennedy, 1999); and (Liu et al., 2007a); and mutation operators (Liu et al., 2007b); and (Andrews, 2006). Many of these techniques have had varying levels of success on the traditional PSO algorithm. It is expected, that these same strategies can be adapted to the combinatorial PSO. Future work aims to investigate the potential to implement these algorithmic improvements to the combinatorial PSO and solve for larger scale combinatorial optimisation problems.

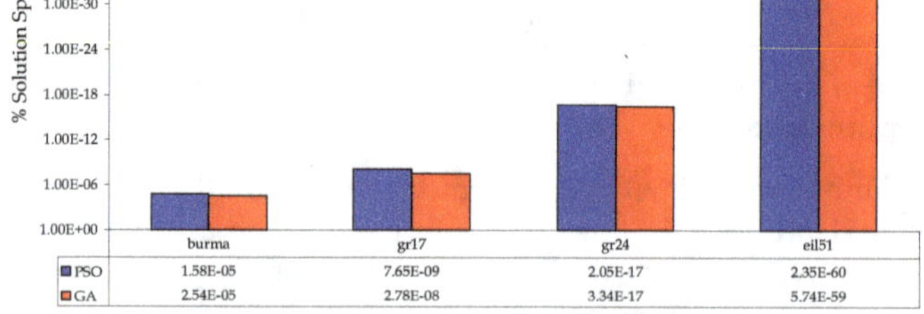

Figure 3. Comparison of the solution space searched by the combinatorial PSO and the GA

9. Conclusion

The PSO's simplicity, robustness, and low computational costs, makes it an ideal method for continuous optimisation problems. Previous efforts to adapt the traditional PSO algorithm to combinatorial spaces have shown varying levels of success. In this chapter, a new combinatorial optimisation PSO that builds on previous works is introduced. A distance metric was introduced to define a metric space for the combinatorial optimisation problem and a syntactic swap operator introduced. Motion was induced by associating a probability sampling function to the velocity profile of a particle on the combinatorial space and invoking the defined swap operator. The proposed algorithm was tested on several instances of TSPLIB and compared to the performance of a GA. Preliminary test results demonstrated superior performance over the GA in all test cases. For larger set sizes, the proposed algorithm failed to converge to the optimal solution. Examination of the sampled solution space suggested that the proposed algorithm suffered from the same rapid convergence and stagnation issues observed in traditional PSO. Further research is needed to clarify the effect of the various tuning parameters on the performance of the proposed algorithm, and their influence on loss of solution diversity. The generalised approach to the algorithm's development allows for the consideration of other metrics on discrete spaces, and the implementation of further algorithmic improvements. Future work aims to investigate methods to mitigate the stagnation issues of the proposed algorithm and extending the combinatorial optimisation PSO's capabilities to other discrete optimisation problems.

10. References

Aardal, K., Hoesel, S. v., Lenstra, J. K. and Stougie, L. (1997). *A Decade of Combinatorial Optimization.* Department of Information and Computing Sciences, Utrecht University, UU-CS-1997-12,

Aguilar, J. (2001). A General Ant Colony Model to solve Combinatorial Optimization Problems. *Revista Colombiana De Computación,* 2, 7 - 18.

Andrews, P. S. (2006). An Investigation into Mutation Operators for Particle Swarm Optimization, *Proceedings of IEEE Congress on Evolutionary Computation,* pp. 1044 - 1051, Vancouver BC, Canada, July,

Applegate, D. L., Bixby, R. E., Chvátal, V. e. and Cook, W. J. (2006). *The Traveling Salesman Problem,* Princeton University Press, New Jersey.

Bean, J. (1994). Genetic Algorithms and Random Keys for Sequencing and Optimization. *ORSA Journal on Computing,* 6, 154 - 160.

Bóna, M. (2004). *Combinatorics of Permutations,* Chapman & Hall/CRC, New York.

Bonyadi, M. R., Azghadi, S. M. R. and Hosseini, H. S. (2007). Solving Traveling Salesman Problem Using Combinational Evolutionary Algorithm, In *Artificial Intelligence and Innovations 2007: From Theory to Applications,* Vol. 247 (Eds, Boukis, C., Pnevmatikakis, L. and Polymenakos, L.) Springer, Boston, pp. 37 - 44.

Cao, Y. J. and Wu, Q. H. (1997). Convergence Analysis of Adaptive Genetic Algorithms, *Proceedings of Genetic Algorithms in Engineering Systems: Innovations and Applications,* pp. 85 - 89, 2 - 4 September,

Clerc, M. and Kennedy, J. (2002). The Particle Swarm - Explosion, Stability, and Convergence in a Multidimensional Complex Space. *IEEE Transactions on Evolutionary Computation*, 6, 58 - 73.

Eberhart, R. C. and Shi, Y. (2000). Comparing Inertia Weights and Constriction Factors in Particle Swarn Optimization, *Proceedings of IEEE Congress on Evolutionary Computation*, pp. 84-88, San Diego, CA,

Festa, P. and Resende, M. G. C. (2008). *Hybrid Grasp Heuristics*. AT&T Labs Research, Florham Park, July

Fieldsend, J. E. and Singh, S. (2002). A Multi-Objective Algorithm Based Upon Particle Swarm Optimisation, an Efficient Data Structure and Turbulence, *Proceedings of UK Workshop on Computational Intelligence*, pp. 37 - 44, UK,

Godsil, C. and Royle, G. (2001). *Algebraic Graph Theory*, Springer-Verlag, New York.

Goldberg, D. and Lingle, J. R. (1985). Alleles, Loci and the TSP, *Proceedings of First International Conference on Genetic Algorithms and Their Applications*, pp. 154 - 159, Hillsdale, Hew Jersey,

Greenwood, G. W. and Zhu, Q. J. (2001). Convergence in Evolutionary Programs with Self-Adaptation. *Evolutionary Computation*, 2, 147 - 157.

He, S., Wu, Q. H., Wen, J. Y., Saunders, J. R. and Paton, R. C. (2004). A Particle Swarm Optimizer with Passive Congregation. *Biosystems*, 78, 135 - 147.

Kadirkamanathan, V., Selvarajah, K. and Fleming, P. J. (2006). Stability Analysis of the Particle Dynamics in Particle Swarm Optimizer. *IEEE Transactions on Evolutionary Computation*, 10, 245 - 255.

Kennedy, J. (1999). Small Worlds and Mega-Minds: Effects of Neighborhood Topology on Particle Swarm Performance, *Proceedings of Congress on Evolutionary Computation*, pp. 1931 - 1938, Washington DC, USA,

Kennedy, J. and Eberhart, R. (1995). Particle Swarm Optimization, *Proceedings of IEEE International Conference on Neural Networks*, pp. 1942-1948, 27th November - 1 December, 1995,

Kennedy, J. and Eberhart, R. C. (1997). A Discrete Binary Version of the Particle Swarm Algorithm, *Proceedings of IEEE International Conference on Systems, Man, and Cybernetics*, pp.

Knuth, D. E. (1998). *The Art of Computer Programming*, Addison-Wesley, Reading.

Liu, H., Abraham, A. and Grosan, C. (2007a). A Novel Variable Neighborhood Particle Swarm Optimization for Multi-Objective Flexible Job-Shop Scheduling Problems, *Proceedings of 2nd International Conference on Digital Information Management*, pp. 138 - 145, October,

Liu, J., Fan, X. and Qu, Z. (2007b). An Improved Particle Swarm Optimization with Mutation Based on Similarity, *Proceedings of Third International Conference on Natural Computation*, pp. Haikou, China,

Martínez García, F. J. and Moreno Pérez, J. A. (2008). *Jumping Frogs Optimization: A New Swarm Method for Discrete Optimization*. Department of Statistics, O. R. and Computing, University of La Laguna, Tenerife, Spain, DEIOC 3/2008,

Misevicius, A., Blažauskas, T., Blonskis, J. and Smolinskas, J. (2004). An Overview of Some Heuristic Algorithms for Combinatorial Optimization Problems. *Information Technology and Control*, 30, 21 - 31.

Moraglio, A. and Poli, R. (2004). Topological Interpretation of Crossover, In *Genetic and Evolutionary Computation - GECCO 2004*, Vol. 3102/2004 Springer Berlin/Heidelberg, Berlin/Heidelberg, pp. 1377 - 1388.

Moreau, L. (2004). Stability of Continuous-Time Distributed Consensus Algorithms, *Proceedings of 43rd IEEE Conference on Decision and Control*, pp. 3998 - 4003, Atlantis, Paradise Island, Bahamas, December,

Olfati-Saber, R. and Murray, R. M. (2003). Consensus Protocols for Networks of Dynamic Agents, *Proceedings of American Control Conference*, pp. 951-956,

Pang, W., Wang, K.-p., Zhou, C.-g. and Dong, L.-j. (2004). Fuzzy Discrete Particle Swarm Optimization for Solving Traveling Salesman Problem, *Proceedings of 4th International Conference on Computer and Information Technology (CIT04), IEEE Computer Society*, pp.

Passino, K. M. (2002). Biomimicry of Bacterial Foraging for Distributed Optimization and Control. *IEEE Control Systems Magazine*, 22, 52 - 67.

Poli, R. and Langdon, W. B. (2007). Markov Chain Models of Bare-Bones Particle Swarm Optimizers, *Proceedings of 9th Annual Conference on Genetic and Evolutionary Computation*, pp. 142 - 149, London, England, 7 - 11 July,

Poli, R., Langdon, W. B., Clerc, M. and Stephens, C. R. (2007). Continuous Optimisation Theory Made Easy? Finite-Element Models of Evolutionary Strategies, Genetic Algorithms and Particles Swarm Optimizers, In *Foundations of Genetic Algorithms*, Vol. 4436/2007 Springer Berlin/Heidelberg, Berlin, pp. 165 - 193.

Reinelt, G. (1991). TSPLIB - Traveling Salesman Problem Library. *ORSA Journal on Computing*, 3, 376 - 384.

Ronald, S. (1997). Distance Functions for Order-Based Encodings, *Proceedings of IEEE International Conference on Evolutionary Computation*, pp. 49 - 54, Indianapolis, USA,

Ronald, S. (1998). More Distance Functions for Order-Based Encodings, *Proceedings of IEEE Conference on Evolutionary Computation*, pp. 558 - 563, IEEE Press

Rudolph, G. (1994). Convergence Analysis of Canonical Genetic Algorithms. *IEEE Transactions on Neural Networks*, 5, 96 - 101.

Rudolph, G. (1996). Convergence of Evolutionary Algorithms in General Search Spaces, *Proceedings of IEEE International Conference on Evolutionary Computation*, pp. 50 - 54, Nagoya, Japan, 20 - 22 May,

Shi, X. H., Zhou, Y., Wang, L. M., Wang, Q. X. and Liang, Y. C. (2006). A Discrete Particle Swarm Optimization Algorithm for Travelling Salesman Problem, In *Computational Methods*(Eds, Liu, G. R., Tan, V. B. C. and Han, X.) Springer Netherlands, Netherlands, pp. 1063 - 1068.

Shi, Y. and Eberhart, R. C. (1998a). A Modified Particle Swarm Optimiser, *Proceedings of IEEE International Conference on Evolutionary Computation*, pp. Anchorage, Alaska, May,

Shi, Y. and Eberhart, R. C. (1998b). Parameter Selection in Particle Swarm Optimization, *Proceedings of 7th International Conference on Evolutionary Programming*, pp. 591 - 600,

Song, X., Chang, C. and Cao, Y. (2008). New Particle Swarm Algorithm for Job Shop Scheduling Problems, *Proceedings of 7th World Congress on Intelligent Control and Automation*, pp. 3996 - 4001, Chongqing, China, 25 - 27 June,

Tasgetiren, M. F., Sevkli, M., Liang, Y. C. and Gencyilmaz, G. (2004). Particle Swarm Optimization Algorithm for Permutation Flowshop Sequencing Problem, In *Ant Colony, Optimization and Swarm Intelligence,* Vol. 3172/2004 Springer Berlin/Heidelberg, Berlin, pp. 382 - 389.

Trelea, I. C. (2003). The Particle Swarm Optimization Algorithm: Convergence Analysis and Parameter Selection. *Information Processing Letters,* 85, 317 - 325.

Wang, K.-p., Huang, L., Zhou, C. G. and Pang, W. (2003). Particle Swarm Optimization for Traveling Salesman Problem, *Proceedings of Second International Conference on Machine Learning and Cybernetics,* pp. 1583 - 1585, Xi'an, China, November,

Zhang, L.-p., Yu, H.-j. and Hu, S.-X. (2005). Optimal Choice of Parameters for Particle Swarm Optimization. *Journal of Zhejiang University Science,* 6, 528-534.

Zheng, Y.-L., Ma, L.-H., Zhang, L.-Y. and Qian, J.-X. (2003). On the Convergence Analysis and Parameter Selection in Particle Swarm Optimization, *Proceedings of Second International Conference on Machine Learning and Cybernetics,* pp. 1802 - 1807, Xi'an, China, November,

13

Discrete Particle Swarm Optimization Algorithm for Flowshop Scheduling

S.G. Ponnambalam[1], N. Jawahar[2] and S. Chandrasekaran[3]
[1]*Monash University,* [2]*Thiagarajar College of Engineering*
[3]*S R M V Polytechnic College*
[1]*Malaysia,* [2,3]*India*

1. Introduction

In the context of manufacturing systems, scheduling refers to allocation of resources over time to perform a set of operations. Manufacturing systems scheduling has many applications ranging from manufacturing, computer processing, transportation, communication, health care, space exploration, education, distribution networks, etc. Scheduling is a process by which limited resources are allocated over time among parallel or sequential activities. Solving such a problem amounts to making discrete choices such that an optimal solution is found among a finite or a countably infinite number of alternatives. Such problems are called combinatorial optimization problems. Typically, the task is complex, limiting the practical utility of combinatorial, mathematical programming and other analytical methods in solving scheduling problems effectively. Manufacturing system entails the acquisition and allocation of limited resources to production activities so as to reduce the manufacturing cycle time and in-process inventory and to satisfy customer demand in specified time. Successful achievement of these objectives lies in efficient scheduling of the system. Scheduling plays an important role in shop floor planning. A schedule shows the planned time when processing of a specific job will start on a machine. It also indicates when a job will get completed on a machine. Scheduling is a decision-making process of sequencing a set of operations on different machines in a manufacturing unit. The objective of scheduling is generally to improve the utilization of resources and profitability of production lines. Scheduling problem is characterized by three components namely:

1. Number of machines, number of jobs and the processing time for each job using appropriate machine
2. A set of constraints such as operation precedence constraint for a given job and operation non-overlapping constraint for a given machine
3. A target function called objective function consisting of single or multiple criteria that must be optimized.

Traditionally, scheduling researchers has shown interest in optimizing a single-objective or performance measure while scheduling, which is not a reality. Practical scheduling problems acquire consideration of several objectives as desired by the scheduler. When multiple criteria are considered, scheduler may wish to generate a schedule which performs better with respect to all the measures under study, such solution does not exist. This chapter presents the application of Discrete Particle Swarm Optimisation Algorithm for solving flowshop scheduling problem (FSP) under single and multiple objective criteria.

2. Flowshop Scheduling

2.1 Description of FSP

In discrete parts manufacturing industries, jobs with multiple operations use machines in the same order. In such case, machines are installed in series. Raw materials initially enter the first machine and when a job has finished its processing on the first machine, it goes to the next machine. When the next machine is not immediately available, job has to wait till the machine becomes available for processing. Such a manufacturing system is called a flowshop, where the machines are arranged in the order in which operations are to be performed on jobs. A flowshop is a conventional manufacturing system where machines are arranged in the order of performing operations on jobs. The technological order, in which the jobs are processed on different machines, is unidirectional. In a flowshop, a job i with a set of m operations $i_1, i_2, i_3, ..., i_m$ is to be completed in a predetermined sequence. In short, each operation except the first has exactly one direct predecessor and each operation except the last one has exactly one direct successor as shown in Figure 1. Thus each job requires a specific immutable sequence of operations to be carried out for it to be complete. This type of structure is sometimes referred as linear precedence structure (Baker, 1974). Further, once started, an operation on a machine cannot be interrupted.

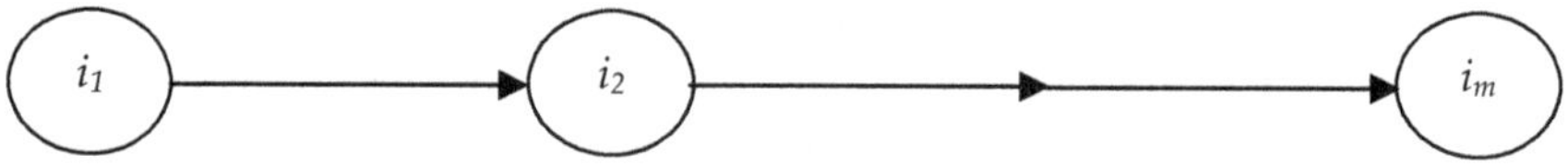

Figure 1. Work Flow in Flowshop

2.2 Characteristics of FSP

Flowshop consists of m machines and there are n different jobs to be optimally sequenced through these machines. The common assumptions used in modelling the flowshop problems are as follows:

- All n jobs are available for processing at time zero and each job follows identical routing through the machines.
- Unlimited storage exists between the machines. Each job requires m operations and each operation requires a different machine.
- Every machine processes only one job at a time and every job is processed by one machine at a time.
- Setup-times for the operations are sequence-independent and are included in processing times.
- The machines are continuously available.
- Individual operations cannot be pre-empted.

Further it is assumed that:

- Each job must be processed to completion.
- In-process inventory is allowed when necessary.

- There is only one machine of each type in the shop.
- Machines are available throughout the scheduling period.
- There is no randomness and the scheduling problem under study is a deterministic scheduling problem. In particular
    - The number of jobs is known and fixed.
    - The number of machines is known and fixed.
    - The processing times are known and fixed, and
    - All other quantities needed to define a particular problem are known and fixed.

The general structure of typical n job m machine FSP is shown in Figure 2.

Job	Processing order		
	M_1	M_2	M_3 ....M_m
J_1	Pt_1	Pt_2	Pt_3 ....Pt_m
J_2	Pt_1	Pt_2	Pt_3 ....Pt_m
J_3	Pt_1	Pt_2	Pt_3 ....Pt_m
..	..	..	.. ..
..	..	..	.. ..
J_n	Pt_1	Pt_2	Pt_3 ....Pt_m

where Pt - processing time of job J in machine M

Figure 2. General Structure of Flowshop

2.3 Solution approaches for FSP

Computational complexity of a problem is the maximum number of computational steps needed to obtain an optimal solution. For example if there are n jobs and m available machines, the available number of schedule to be evaluated to get an optimal solution is $(n!)^m$. In a permutation flow based manufacturing system, the number of available schedules is $n!$. Based on the complexity of the problem, all problems can be classified into two classes, called P and NP in the literature. Class P consists of problems for which the execution time of the solution algorithms grows polynomially with the size of the problem. Thus, a problem of size m would be solvable in time proportional to m^k, when k is an exponent. The time taken to solve a problem belonging to the NP class grows exponentially, thus this time would grow in proportion to t^m, where t is some constant. In practice, algorithms for which the execution time grows polynomially are preferred. However, a widely held conjecture of modern mathematics is that there are problems in NP class for which algorithms with polynomial time complexity will never be found (French, 1982). These problems are classified as $NP-hard$ problems. Unfortunately, most of the practical scheduling problems belong to the $NP-hard$ class (Rinnooy Kan, 1976). Many scheduling problems are polynomially solvable, or NP-hard in that it is impossible to find an optimal solution here without the use of an essentially enumerative algorithm. FSP is a widely researched combinatorial optimization problem, for which the computational effort increases exponentially with problem size (Jiyin Liu & Colin Reeves, 2001; Brucker, 1998; Sridhar & Rajendran, 1996; French, 1982). In FSP, the computational complexity increases with increase in problem size due to increase in number of jobs and/or number of machines.

To find exact solution for such combinatorial problems, a branch and bound or dynamic programming algorithm is often used when the problem size is small. Exact solution methods are impractical for solving FSP with large number of jobs and/or machines. For the large-sized problems, application of heuristic procedures provides simple and quick method of finding best solutions for the FSP instead of finding optimal solutions. A heuristic is a technique which seeks (and hopefully finds) good solutions at a reasonable computational cost. A heuristic is approximate in the sense that it provides a good solution for relatively little effort, but it does not guarantee optimally. A heuristic can be a rule of thumb that is used to guide one's action. Heuristics for the FSP can be a constructive heuristics or improvement heuristics. Various constructive heuristics methods have been proposed by Johnson, 1954; Palmer, 1965; Campbell et al., 1970; Dannenbring 1977 and Nawaz et al. 1983. Literature shows that constructive heuristic methods give very good results for NP-hard combinatorial optimization problems. This builds a feasible schedule from scratch and the improvement heuristics try to improve a previously generated schedule by applying some form of specific improvement methods. An application of heuristics provides simple and quick method of finding best solutions for the FSPs instead of finding optimal solutions (Ruiz & Maroto, 2005; Dudek et al. 1992). Johnson's algorithm (1954) is the earliest known heuristic for the FSP, which provides an optimal solution for two-machine problem to minimize makespan. Palmer (1965) developed a very simple heuristic in which for every job a "slope index" is calculated and then the jobs are scheduled by non-increasing order of this index. Ignall & Schrage (1965) applied the branch and bound technique to the flowshop sequencing problem. Campbell et al. (1970) developed a heuristic algorithm known as CDS algorithm and it builds $m-1$ schedules by clustering the m original machines into two virtual machines and solving the generated two machine problem by repeatedly using Johnson's rule. Dannenbring's (1977) Rapid Access heuristic is a mixture of the previous ideas of Johnson's algorithm and Palmer's slope index. Nawaz et al.'s (1983) NEH heuristic is based on the idea that jobs with high processing times on all the machines should be scheduled as early in the sequence as possible. NEH heuristics seems to be the performing better compared to others. Heuristic algorithms are conspicuously preferable in practical applications. Among the most studied heuristics are those based on applying some sort of greediness or applying priority based procedures including, e.g., insertion and dispatching rules. The main drawback of these approaches, their inability to continue the search upon becoming trapped in a local optimum, leads to consideration of techniques for guiding known heuristics to overcome local optimality (Jose Framinan et al. 2003). And also the heuristics has the problems like

1. Lack of comprehensiveness
2. Little robustness of conclusions
3. Weak/partial experimental design

For these reasons, one can investigate the application of metaheuristic search methods for solving optimization problems. It is a set of algorithmic concepts that can be used to define heuristic methods applicable to wide set of varied problems. The use of metaheuristics has significantly produced good quality solutions to hard combinatorial problems in a reasonable time. It is defined as an iterative generation process which guides a subordinate heuristic by combining intelligently different concepts for exploring and exploiting the search space, learning strategies are used to structure information in order to find efficiently near-optimal solutions (Osman & Laporte, 1996). The fundamental properties which characterize metaheuristics are as follows (Christian Blum & Andrea Roli, 2003):

- The goal is to efficiently explore the search space in order to find (near-) optimal solutions.
- Techniques which constitute metaheuristic algorithms range from simple local search procedures to complex learning processes.
- Metaheuristic algorithms are approximate and usually non-deterministic.
- They may incorporate mechanisms to avoid getting trapped in confined areas of the search space.
- The basic concepts of metaheuristics permit an abstract level description.
- Metaheuristics are not problem-specific.
- Metaheuristics may make use of domain-specific knowledge in the form of heuristics that are controlled by the upper level strategy.
- Today's more advanced metaheuristics use search experience (embodied in some form of memory) to guide the search.

Metaheuristics or Improvement heuristics are extensively employed by researchers to solve scheduling problems (Chandrasekaran et al. 2006; Suresh & Mohanasundaram, 2004; Hisao Ishibuchi et al. 2003; Lixin Tang & Jiyin Liu, 2002; Eberhart & Kennedy, 1995). Improvement methods such as Genetic Algorithm (Chan et al. 2005; Ruiz et al. 2004; Sridhar & Rajendran, 1996), Simulated Annealing algorithm (Ogbu & Smith, 1990), Tabu Search algorithm (Moccellin & Nagamo, 1998) and Particle Swarm Optimization algorithm (Rameshkumar et al. 2005; Prabhaharan et al. 2005; Faith Tasgetiren et al. 2004) have been widely used by researchers to solve FSPs. Metaheuristic algorithms such as Simulated Annealing (SA) and Tabu Search (TS) methods are single point local search procedures where, a single solution is improved continuously by an improvement procedure. Algorithms such as Genetic Algorithm (GA), Ant Colony Optimization (ACO) algorithm and Particle Swarm Optimization (PSO) algorithm belongs to population based search algorithms. These are designed to maintain a set of solution transiting from a generation to the next. The family of metaheuristics includes, but is not limited to, GA, SA, ACO, TS, PSO, evolutionary methods, and their hybrids.

2.4 Performance measures considered

Measures of schedule performance are usually functions of the set of completion times in a schedule. Performance measures can be classified as regular and non-regular. A regular measure is one in which the penalty function is non-decreasing in terms of job completion times. Some examples of regular performance measures are makespan, mean flowtime, total flowtime, and number of tardy jobs. Performance measures, which are not regular, are termed non-regular. That is, such measures are not an increasing function with respect to job completion times. Some examples of non-regular measures are earliness, tardiness, and completion time variance. In this chapter, the performance measures namely minimization of makespan, total flowtime and completion time variance is considered for solving flowshop scheduling problems. Makespan (C_{max}) has been considered by many scheduling researchers (Ignall & Scharge, 1965; Campbell et al. 1970; Nawaz et al.1983; Framinan et al. 2002; Ruiz & Maroto, 2005). Makespan is defined as the time required for processing all the jobs or the maximum time required for completing a given set of jobs. Minimization of makespan ensures better utilization of the machines and leads to a high throughput (Framinan et al. 2002; Ruiz & Maroto, 2005). Makespan is computed using equation (1).

$$C_{max} = \max\{C_i, i = 1,2,........,n\} \quad (1)$$

The time spend by a job in the system has been defined as its flow time. Total flowtime is defined as the sum of completion time of every job or total time taken by all the jobs. Total flowtime (F_Σ) of the schedule is computed using equation (2). Minimizing total flowtime results in minimum work-in-process inventory (Chandrasekharan Rajendran & Hans Ziegler, 2005).

$$F_\Sigma = \sum_{i=1}^{n} C_i \quad (2)$$

Completion time variance is defined as the variance about the mean flowtime and is computed using equation (3). Minimizing completion time variance (V_T) serves to minimize variations in resource consumption and utilization (Gowrishankar et al. 2001; Gajpal & Rajendran, 2006; Viswanath Kumar Ganesan et al. 2006).

$$V_T = \frac{1}{n}\sum_{i=1}^{n}(C_i - \overline{F})^2 \quad (3)$$

where $\overline{F} = \frac{F_T}{n}$ is the mean flowtime.

3. Particle Swarm Optimization Algorithm

3.1 Features of PSO

Particle Swarm Optimization (PSO) algorithm is an evolutionary computation technique developed by Eberhart & Kennedy in 1995 inspired by social behavior of bird flocking or fish schooling. PSO is a stochastic, population-based approach for solving problems (Kennedy & Eberhart, 1995). It is a kind of swarm intelligence that is based on social-psychological principles and provides insights into social behavior, as well as contributing to engineering applications. PSO algorithm has been successfully used to solve many difficult combinatorial optimization problems. PSO algorithm is problem-independent, which means little specific knowledge relevant to a given problem is required. All we have to know is the fitness evaluation of each solution. This advantage makes PSO more robust than many search algorithms. In the last couple of years the particle swarm optimization algorithm has reached the level of maturity necessary to be interesting from an engineering point of view. It is a potent alternative optimizer for complex problems and possesses many attractive features such as:

- Ease of implementation: The PSO is implemented with just a few lines of code, using only basic mathematical operations.
- Flexibility: Often no major adjustments have to be made when adapting the PSO to a new problem.
- Robustness: The solutions of the PSO are almost independent of the initialization of the swarm. Additionally, very few parameters have to be tuned to obtain quality solutions.

- Possibility to combine discrete and continuous variables. Although some authors present this as a special feature of the PSO (Sensarma et al., 2002), others point out that there are potential dangers associated with the relaxation process necessary for handling the discrete variables (Abido, 2002). Simple round-off calculations may lead to significant errors.
- Possibility to easily tune the balance between local and global exploration.
- Parallelism: The PSO is inherently well suited for parallel computing. The swarm population can be divided between many processors to reduce computation time.

3.2 Applications of PSO

In recent years, PSO has been successfully applied in many areas. Currently, PSO has been implemented in a wide range of research areas such as functional optimization, pattern recognition, neural network training, fuzzy system control etc. and obtained significant success. PSO is widely applied and focused by researchers due to its profound intelligence background and simple algorithm structure. Many proposals indicate that PSO is relatively more capable for global exploration and converges more quickly than many other heuristic algorithms. It solves a variety of optimization problems in a faster and cheaper way than the evolutionary algorithms in the early iterations. One of the reasons that PSO is attractive is that there are very few parameters to adjust. One version, with very slight variation (or none at all) works well in a wide variety of applications. PSO has been used for approaches that can be used across a wide rage of applications, as well as for specific applications focused on a specific requirement. PSO has been applied to the analysis of human tremor. The diagnosis of human tremor, including Parkinson's disease and essential tremor, is a very challenging area. PSO has been used to evolve a neural network that distinguishes between normal subjects and those with tremor. Inputs to the network are normalized movement amplitudes obtained from an actigraph system. The method is fast and accurate (Eberhart & Hu, 1999). While development of computer numerically controlled machine tools has significantly improved productivity, there operation is far from optimized. None of the methods previously developed is sufficiently general to be applied in numerous situations with high accuracy. A new and successful approach involves using artificial neural networks for process simulation and PSO for multi-dimensional optimization. The application was implanted using computer-aided design and computer-aided manufacturing (CAD/CAM) and other standard engineering development tools as the platform (Tandon, 2000). Another application is the use of particle swarm optimization for reactive power and voltage control by a Japanese electric utility (Yoshida et al., 1999). PSO has also been used in conjunction with a back propagation algorithm to train a neural network as a state-of-charge estimator for a battery pack for electric vehicle use. Determination of the battery pack state of charge is an important issue in the development of electric and hybrid / electric vehicle technology. The state of charge is basically the fuel gauge of an electric vehicle. A strategy was developed to train the neural network based on a combination of particle swarm optimization and the back propagation algorithm. Finally, one of the most exciting applications of PSO is that by a major American corporation to ingredient mix optimization. In this work, "ingredient mix" refers to the mixture of ingredients that are used to grow production strains of microorganisms that naturally secrete of manufacture something of interest. Here, PSO was used in parallel with traditional industrial optimization methods. PSO provided an optimized ingredient mix that provided over twice the fitness as the mix found using traditional methods, at a very different location in ingredient space. PSO was shown to be robust: the occurrence of an ingredient becoming contaminated hampered the search for a few iterations but in the end did not result in poor final results. PSO, by its nature, searched a much larger portion of the problem space than the traditional method. Generally speaking, particle swarm optimization, like the other evolutionary computation algorithms, can be applied to solve most optimization problems and problems that can be converted to optimization problems. Among the application areas with the most potential are system design, multi-objective optimization, classification, pattern recognition, biological system modelling, scheduling (planning), signal processing, games, robotic applications, decision making, simulation and identification. Examples include fuzzy controller design, job shop scheduling, real time robot path planning, image segmentation, EEG signal simulation, speaker verification, time-frequency analysis, modelling of the spread of antibiotic resistance, burn diagnosing, gesture recognition and automatic target detection, to name a few (Eberhart & Shi, 2001).

3.3 Working of PSO

PSO is initialized with a swarm of random feasible solutions and searches for optima by updating velocities and positions. PSO algorithm is initialized with a set of several random particles called a swarm. A set of moving particles (the swarm) is initially thrown inside the multi-dimensional search space. Each particle is a potential solution, which has the ability to remember its previous best position and current position, and it survives from generation to generation. Each particle has the following features:

- It has a position and a velocity
- It knows its neighbours, best previous position and objective function value.
- It remembers its best previous position.

At each time step, the behavior of a given particle is a compromise between three possible choices

- To follow its own way
- To go towards its best previous position
- To go towards the best neighbour's best previous position, or forwards the best neighbour.

The swarm is typically modelled by particles in multi-dimensional space that have a position and a velocity. These particles fly through hyperspace and have two essential reasoning capabilities: their memory of their own best position and knowledge of their neighborhood's best, "best" simply meaning the position with the smallest objective value. Members of a swarm communicate good positions to each other and adjust their own position and velocity based on these good positions. PSO shares many similarities with evolutionary computation techniques such as GA, SA, TS and ACO algorithms. The PSO system is initialized with a swarm of random solutions and searches for optima by updating generations. The advantages of PSO are that PSO is easy to implement and there are few parameters to adjust. PSO has been successfully applied in many areas: function optimization, artificial neural network training, fuzzy system control, and other areas where GA can be applied. Most of evolutionary techniques have the following procedure:

1. Random generation of an initial population
2. Reckoning of a fitness value for each subject. It will directly depend on the distance to the optimum.

3. Reproduction of the population based on fitness values.
4. If requirements are met, then stop. Otherwise go back to step 2.

Evolutionary Algorithms use a population of potential solutions (points) of the search space. These solutions (initially randomly generated) are evolved using different specific operators which are inspired from biology. Through cooperation and competition among the potential solutions, these techniques often can find near-optimal solutions quickly when applied to complex optimization problems. There are some similarities between PSO and Evolutionary Algorithms:

1. Both techniques use a population (which is called *swarm* in the PSO case) of solutions from the search space which are initially random generated;
2. Solutions belonging to the same population interact with each other during the search process;
3. Solutions are evolved using techniques inspired from the real world.

PSO shares many common points with GA. Both algorithms start with a group of a randomly generated population; both have fitness values to evaluate the population. Both update the population and search for the optimum with random techniques. Both systems do not guarantee success. However, PSO does not have genetic operators like crossover and mutation. Particles update themselves with the internal velocity. The information sharing mechanism in PSO is significantly different. In GA, chromosomes share information with each other. So the whole population moves like one group towards an optimal area. In PSO, only global or local best particle gives out the information to others. It is a one-way information sharing mechanism. Compared with GA, all the particles tend to converge to the best solution quickly even in the local version in most cases/ PSO optimization algorithm uses a set of particles called a swarm, similar to chromosomes in a binary-coded Genetic Algorithm (GA). PSO and ACO are optimization algorithms based on the behavior of swarms (birds, fishes) and ants respectfully. However, the particles are multidimensional points in real space during the optimization. The PSO optimization run starts with a user-specified swarm size and objective function used to evaluate objection function values, called fitness in GA terminology. The particles are initialized randomly within the variable bounds and they search for the optimum (maximum or minimum) in the search space with some communication between particles. For a maximization (or minimization) problem, the particles will move towards the particle with the highest (or least) objective function value using a position update equation, that is stochastic. This is how randomness in introduced to PSO algorithm. This position update method is similar to the use of crossover and mutation operations used to generate new individuals in a new generation in the GA. However, the PSO differs in that, updates of particle position usually involve the best particles (global or in the neighborhood) of each particle. The position updating tends to always exploit the best solution found so far. While this may lead to premature convergence, when all particles positions become equal to that of the best particle (i.e., no diversity), there are schemes designed to prevent such premature convergence. In the PSO literature, several neighborhood schemes have been developed for the particle updating (Merkle and Middendorf, 2000). This chapter aims to develop a metaheuristic algorithm called PSO algorithm which is suitable for solving FSPs with the objective of minimising three performance measures namely makespan, total flowtime and completion time variance. Firstly, a single objective PSO is proposed and the above performance measures are considered individually. Performance of the proposed single objective PSO is tested by solving a large set of benchmark FSPs available in the literature having number of jobs varying from 5 to 500 and number of machines from 5 to 20.

3.4 Structure of PSO Algorithm

The pseudo-code of the simple PSO algorithm and its general framework are given in Figures 3 and 4 respectively.

The basic elements of PSO algorithm are summarized below:

Particle: X_i^t denotes the ith particle in the swarm at iteration t and is represented by n number of dimensions as $X_i^t = [x_{i1}^t, x_{i2}^t, \ldots, x_{in}^t]$, where x_{ij}^t is the position value of the ith particle with respect to the jth dimension ($j = 1, 2, \ldots, n$).

Population: pop^t is the set of NP particles in the swarm at iteration t, i.e., $pop^t = [X_1^t, X_2^t, \ldots, X_{NP}^t]$.

Sequence: We introduce a new variable π_i^t, which is a permutation of jobs implied by the particle X_i^t. It can be described as $\pi_i^t = [\pi_{i1}^t, \pi_{i2}^t, \ldots, \pi_{in}^t]$, where π_{ij}^t is the assignment of job j of the particle i in the permutation at iteration t.

```
Initialize swarm
Initialize velocity
Initialize position
Initialize parameters
      Evaluate particles
      Find the local best
      Find the global best
Do
{
            Update velocity
            Update position
            Evaluate
            Update local best
            Update global best
} ( until termination)
```

Figure 3. Pseudocode of the PSO Algorithm

Particle velocity: V_i^t is the velocity of particle i at iteration t. It can be defined as $V_i^t = [v_{i1}^t, v_{i2}^t, \ldots, v_{in}^t]$, where v_{ij}^t is the velocity of particle i at iteration t with respect to the jth dimension.

Local best: P_i^t represents the best position of the particle i with the best fitness until iteration t, so the best position associated with the best fitness value of the particle i obtained so far is called the *local best*. For each particle in the swarm, P_i^t can be determined and updated at each iteration t. In a minimization problem with the objective function $f(\pi_i^t)$ where π_i^t is the corresponding sequence of particle X_i^t, the local best P_i^t of the ith particle is obtained such that $f(\pi_i^t) \leq f(\pi_i^{t-1})$ where π_i^t is the corresponding permutation of local best P_i^t

and π_i^{t-1} is the corresponding sequence of local best P_i^{t-1}. To simplify, we denote the fitness function of the local best as $f_i^{pb} = f(\pi_i^t)$. For each particle, the local best is defined as $P_i^t = [p_{i1}^t, p_{i2}^t, ..., p_{in}^t]$ where p_{ij}^t is the position value of the ith local best with respect to the jth dimension ($j = 1,2,...,n$).

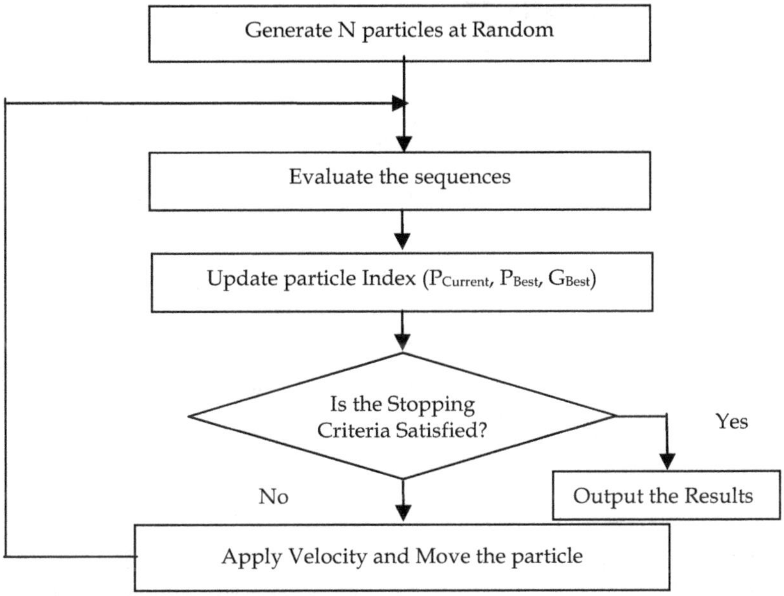

Figure 4. The Framework of PSO Algorithm

4. Discrete PSO Algorithm for Single-Objective FSP

4.1 Pseudocode of the proposed discrete PSO algorithm

Particle Swarm Optimization algorithm starts with a population of randomly generated initial solutions called particles (swarm). It is to be noted that the particle structure is taken as a string, which consists of job numbers in certain order. The order of jobs in the string represents a sequence. After the swarm is initialized, each potential solution is assigned a velocity randomly. The length of the velocity of each particle $\|v\|$ is generated randomly between *0* and *n* (Rameshkumar et al. 2005; Chandrasekaran et al. 2006) and the corresponding lists of transpositions $(i_q, j_q); q = 1, \|v_k\|$ are generated randomly for each particle. The above formulation permits exchange of jobs $(i_1, j_1), (i_2, j_2) ...... (i_{\|v\|}, j_{\|v\|})$ in the given order. Each particle keeps track of its improvement and the best objective function value achieved by the individual particles so far is stored as local best solution $({}^eP_k^t)$, and the overall best objective function achieved by all the particles together so far is stored as the global best solution (G_b^t). The particle velocity and position are updated continuously in all iterations. The iterative improvement process is continued afterwards to further improve the solution quality. The Pseudocode of the proposed discrete PSO algorithm is shown in Figure 5.

Initialize swarm P $t=0;$
Initialize velocity v_k^t *and position* P_k^t
Initialize parameters
Evaluate particles
Find the local best ${}^eP_k^t$ *and global best* G_b^t
Do
{
for $(k=1,N)$
Update Velocity v_k^{t+1};
Update Position P_k^{t+1};
Evaluate all particles;
Update ${}^eP_k^{t+1}$ *and* G^{t+1}, $(k=1,N)$;
$t \to t+1$;
} $(\text{while } t < t_{max})$

Figure 5. Pseudocode of the Proposed Discrete PSO Algorithm

The particle velocity and position are continuously updated using equation (4) and (5).

$$v_k^{t+1} = C_1 U_1 v_k^t + C_2 U_2 \,\text{rand}()({}^eP_k^{t+1} - P_k^{t+1}) + C_3 U_3 \,\text{rand}()(G_k - P_k^{t+1}) \tag{4}$$

$$P_k^{t+1} = P_k^t + v_k^{t+1} \tag{5}$$

where C_1, C_2 and C_3 is called acceleration constants. The acceleration constants C_1, C_2 and C_3 in equation (4) guide every particle toward local best and the global best solution during the search process. Low acceleration value results in walking far from the target, namely local best and the global best. High value results in premature convergence of the search process.

4.2 Procedural steps of the Discrete PSO Algorithm

The step by step procedure for implementing the proposed discrete PSO algorithm is as follows.

Step1: Initialize a swarm P_i with random positions and velocities in the problem space X.

Step2: For each particle, evaluate the desired optimization fitness function

Step3: Compare the fitness function with its previous best. If current value is better than previous best, then set previous best equal to current value and P_i equal to the current location X_i.

Step4: Identify the particle in the neighborhood with the best success so far, and assign its index to the variable G.

Step5: Apply local search algorithm to all the particles at the end of each iteration and evaluate for the objective function.

Step6: Change the velocity and position of the particle according to equation (4) and equation (5).

Step7: Loop to step (2) until a criterion is met (usually number of iterations).

4.3 Numerical Illustrations

An example illustrating the process of updating the velocity and the position of a sequence is explained as follows:

Velocity update: The procedure for updating the velocity of all the particles in each iteration is as follows: For example, let us assume

The sequence $P_k^t = \{2,3,4,1\}$; $C_1 = 1,\ C_2 = 2,\ C_3 = 2$; $U_1 = 0.2,\ U_2 = 0.4,\ U_3 = 0.3$; $\|V_k\| = 2$, $v = (1,4),(2,3)$; ${}^eP_k^t = (1,4,3,2)$ and $G_b^t = (3,1,4,2)$.

Velocity of the particle *k* at time step $t+1$ namely V_k^{t+1} is obtained using equation (4)

V_k^{t+1} = 1x 0.2 [(1,4),(2,3)] ⊕ 2 x 0.4 [(1,4,3,2) - (2,3,4,1)] ⊕ 2 x 0.3 [(3,1,4,2) - (2,3,4,1)]

where [(1,4,3,2) - (2,3,4,1)] represents a velocity such that applying the resulting velocity to the current particle (2,3,4,1) yields a position (1,4,3,2).

Thus, V_k^{t+1} = 0.2 [(1,4), (2,3)] ⊕ 0.8 [(2,3), (1,4)] ⊕ 0.6 [(1,2), (1, 4)]

= ((1, 4),(2, 3),(1, 2))

Position update: Position of the particle k at time step $t+1$ namely P_k^{t+1} is obtained using equation (5) by applying V_k^{t+1} over P_k^t as follows.

P_k^{t+1} =(2,3,4,1) + ((1,4), (2,3),(1,2));

= (1,3,4,2) + ((2,3),(1,2)); =(1,4,3,2) + (1,2);

= (4,1,3,2)

4.4 Performance Comparison

An extensive performance analysis using proposed discrete PSO algorithm is carried out by means of evaluating the performance measures by solving the benchmark FSPs of Taillard (1993). Extensive experiments are conducted to fix the parameters like number of particles, number of iterations, selection of learning coefficients and initial swarm generation. The evaluation of proposed discrete PSO algorithm is coded in Linux C and run on an Intel Pentium III 900MHz PC with 128 MB memory.

Number of iterations: Number of iterations or termination criterion is a condition that the search process will be terminated. It might be a maximum number of iteration or maximum CPU times are normally to terminate the search process (Liu & Reeves, 2001; Gowrishankar et al. 2001). In this chapter, for the single-objective optimization problems, an evaluation of *1000 x n x m* number of sequences or particles is taken as the termination criterion.

Number of particles: Experiments have been conducted to identify the optimal swarm size by solving a set of 30 different instances of Taillard (1993) for makespan objective with 20 jobs and machines varying from 5, 10 and 20 using discrete PSO algorithm. In experimentation, the performance of the algorithm is better with swarm size 80 and the same has been used throughout our evaluation.

Learning coefficients: The roll of learning coefficients or acceleration constants, namely C_1, C_2 and C_3 guide every particle towards the local best and the global best solutions during the search process. Low acceleration value results in walking far from the target, namely local best and the global best. High value results in premature convergence of the search process. Experiments have been conducted using different combinations of learning coefficients. To determine the best combinations of C_1, C_2 and C_3 values by solving a set of 30 FSPs for makespan objective with 20 jobs and machines varying from 5, 10 and 20 using

the proposed PSO algorithm. The values $C_1 = 1, C_2 = 2$ and $C_3 = 2$ shows better performance and the same, has been used throughout our study.
Velocity coefficients: The velocity update is carried out after every iteration to improve the search process. The velocity coefficients, namely U_1, U_2 and U_3 guides the search to find the optimal solution quickly. As per the experiments, the values for U_1, U_2 and U_3 are generated randomly between 0 and 1.
Initial Swarm Generation: For the generation of initial swarm one particle is generated from the results obtained by certain algorithms for the desired optimization fitness function and remaining particles of the swarm is constructed in a way that a permutation is produced randomly. The particle generated from certain algorithms is added with randomly generated particles at the beginning of the search. This insertion of the particle in initial swarm is to find better sequences in each iteration of the search. And also it improves the performance of discrete PSO algorithm in terms of finding near-optimal solutions. The algorithms selected for generating the particle for different objective functions are listed below. For makespan objective, one particle is generated using NEH heuristic of Nawaz et al. (1983) and is added to the swarm. For total flowtime objective, one particle is generated based on the heuristic developed by Rajendran. (1993) and is added to the swarm. For completion time variance objective, a particle is generated based on the algorithm developed by Gajpal & Rajendran (2006), and is added to the swarm. These algorithms have better start with the respective objectives. Performance of the proposed discrete PSO with respect to makespan objective is carried out in comparison with the benchmark solutions given by Taillard (1993) and with the results published in the literature. The quality measure namely, "Average Relative Percent Deviation" ($\overline{RPD}$) is considered for the evaluation. During comparison, the corresponding better values reported in the literature are taken. The RPD is computed using equation (6).

$$RPD = [G - C^* / C^*] \times 100 \quad (6)$$

where, G represents the global best solution obtained by the proposed algorithm for a given problem and C^* represents the upper bound value reported in the literature for the corresponding objective function. Some sample results of problems ta001-ta010 of Taillard (1993) is presented in Table 1.

Instances	Problem	Results Reported	Results Obtained	RPD
20 x 5	ta001	1278	1278	0.0000
	ta002	1359	1360	0.0736
	ta003	1081	1088	0.6475
	ta004	1293	1293	0.0000
	ta005	1235	1235	0.0000
	ta006	1195	1195	0.0000
	ta007	1239	1239	0.0000
	ta008	1206	1206	0.0000
	ta009	1230	1237	0.5691
	ta010	1108	1108	0.0000
		$\overline{RPD}$		**0.1290**

Table 1. Sample Results for Makespan

In order to evaluate the performance of the proposed discrete PSO with respect to the total flowtime objective, the results are compared with the results of the popular performing heuristics developed by Liu & Reeves (2001), M-MMAS Algorithm and PACO Algorithm (Rajendran & Ziegler, 2004). Some sample results of problems ta001-ta010 for total flowtime criteria is presented in Table 2.

Instances	Problem	Results Reported	Results Obtained	RPD
20 x 5	ta001	14056 [3]	14033	-0.1636
	ta002	15151 [2]	15151	0.0000
	ta003	13403 [3]	13313	-0.6715
	ta004	15486 [2]	15459	-0.1744
	ta005	13529 [3]	13529	0.0000
	ta006	13123 [3]	13123	0.0000
	ta007	13559 [2]	13548	-0.0811
	ta008	13968 [1]	13948	-0.1432
	ta009	14317 [2]	14315	-0.0140
	ta010	12968 [2]	12943	-0.1928
		$\overline{RPD}$		**-0.1441**

Note: Superscript (1) refers to Heuristic Algorithm (Liu & Reeves, 2001) (2) M-MMAS Algorithm (Rajendran & Ziegler, 2004) (3) PACO Algorithm (Rajendran & Ziegler, 2004)

Table 2. Sample Results for Total Flowtime

Instances	Problem	Results Reported	Results Obtained	RPD
20 x 5	ta001	73040.55 [3]	72060.23	-1.3422
	ta002	90885.27 [2]	89238.17	-1.8123
	ta003	53894.49 [2]	53851.95	-0.0789
	ta004	89822.05 [4]	87104.42	-3.0256
	ta005	72350.55 [2]	72020.43	-0.4563
	ta006	71665.73 [2]	70817.64	-1.1834
	ta007	69088.45 [2]	68367.69	-1.0432
	ta008	70214.31 [2]	69793.85	-0.5988
	ta009	73329.22 [2]	72284.98	-1.4240
	ta010	52580.03 [1]	52015.34	-1.0740
		$\overline{RPD}$		**-1.2039**

Note: Superscript (1) refers to PACO Algorithm (Rajendran & Ziegler, 2004) (2) MMAS Ant Colony Algorithm (Stuetzle, 1998) (3) NACO Algorithm with position-job insertion local search (Gajpal & Rajendran, 2006) (4) NACO Algorithm with job-index based local search (Gajpal & Rajendran, 2006)

Table 3. Sample Results for Completion Time Variance

The performance of the proposed discrete PSO algorithm with respect to completion time variance criterion, the results are compared with the results of ant colony algorithm with random-job insertion local search by Gajpal & Rajendran (2006), M-MMAS Ant Colony Algorithm by Stuetzle(1998), PACO Algorithm by Rajendran & Ziegler(2004), and three

NACO Algorithm with position-job insertion and job-index based local searches by Rajendran & Ziegler (2004). To our knowledge, the results of completion time variance objective using PSO algorithm are not available in literature, the performance of the proposed algorithm is compared with other metaheuristic results. Some sample results of problems ta001-ta010 of Taillard (1993) for completion time variance objective are presented in Table 3.

The results show that the proposed single-objective discrete PSO algorithm performs better. The negative sign in RPD values shows that the proposed discrete PSO algorithm generates better results than the results reported in the literature considered. The summary of $\overline{\text{RPD}}$ values obtained for all the FSP instances of Taillard (1993) are presented in Table 4.

Instances	Number of problems	Makespan	Total Flowtime	Completion Time Variance
20 x 5	10	0.1290238	-0.1440537	-1.2038674
20 x10	10	0.5334462	-0.0164544	-1.7613968
20 x 20	10	0.5329960	-0.0260092	-0.8586390
50 x 5	10	0.0890855	-0.2925054	-0.9330275
50 x 10	10	1.7541958	-0.0108922	-0.2059756
50 x 20	10	2.9814187	0.2434647	1.7126618
100 x 5	10	0.1713382	-0.7238382	1.2988817
100 x 10	10	0.6882989	-0.1191928	0.9198400
100 x 20	10	2.8784086	0.1476830	3.4646301
200 x 10	10	0.5498368	1.8246721	0.0000000
200 x 20	10	2.7011408	1.4120018	0.0000000
500 x 20	10	1.8172343	1.4205378	0.0000000

Table 4. $\overline{\text{RPD}}$ Values Obtained for the Various FSP Instances

The proposed discrete PSO algorithm generates good results with reasonable CPU time. CPU time taken by the proposed discrete PSO algorithm for various FSPs are presented in Table 5.

Instances	Number of Problems	Makespan	Total Flowtime	Completion Time Variance
20x5	10	0m25.164s	0m5.201s	0m6.642s
20x10	10	1m36.844s	0m12.113s	0m33.619s
20x20	10	6m22.854s	0m35.139s	2m16.764s
50x5	10	13m44.973s	0m39.888s	1m10.433s
50x10	10	55m38.305s	1m45.854s	6m19.487s
50x20	10	110m32.087s	10m33.215s	32m41.970s
100x5	10	19m42.310s	4m17.995s	10m39.676s
100x10	10	26m3.295s	9m22.616s	45m1.041s
100x20	10	62m14.918s	33m57.255s	84m4.257s
200x10	10	143m25.161s	41m33.599s	50m27.703s
200x20	10	166m27.657s	79m22.342s	129m58.384s
500x20	10	543m32.695s	792m17.371s	410m50.485s

Table 5. CPU time taken for Various FSP Instances

5. Discrete PSO Algorithm for Multi-Objective FSP

5.1 Concept and terminology

The real-world scheduling problems are multi-objective in nature. In such cases, several objectives must be simultaneously considered when evaluating the quality of the proposed solution. In multi objective decision problems one desires to simultaneously optimize more than one performance objectives such as makespan, tardiness, mean flowtime of jobs, etc. multi-objective optimization usually results in a set of non-dominated solutions instead of a single solution. The goal of multi-objective scheduling is to find a set of compromising schedules satisfying different objectives under consideration. For a given finite set of schedules generated by using a suitable algorithm for a multi-objective scheduling problem, various objective functions $f(x)=\{f_1(x),f_2(x),....,f_k(x)\}$ can be evaluated. These schedules are to be compared and a set of schedules called *non-dominated solutions* are to be identified. For those solutions, no improvement in any objective function is possible without scarifying at least one of the other objective functions. Some researchers have developed multi-objective metaheuristics for solving flowshop scheduling problems (Pasupathy et al. 2006; Prabhaharan et al. 2005; Loukil et al. 2005; Suresh & Mohanasundaram, 2004; Hisao Ishibuchi et al. 2003; Ishibuchi & Murata, 1998; Sridhar & Rajendran, 1996). A survey of multi-objective scheduling problems is given by T'kindt & Billaut (2001). A multi-objective PSO algorithm has been proposed for minimizing weighted sum of makespan and maximum earliness (Prabhaharan et al. 2005). A Pareto archived simulated annealing algorithm for multi-objective scheduling has been proposed (Suresh & Mohanasundaram, 2004). Hisao Ishibuchi et al. (2003) proposed a modified multi-objective genetic local search algorithm (MMOGLS) for multi-objective FSP. They showed that the performance of the evolutionary multi-objective optimization algorithm can be improved by hybridization with local search. They apply multi-objective GA for PFSP and the results are compared with results published in the literature. Pasupathy et al. (2005) proposed a pareto-ranking based multi-objective GA called Pareto genetic algorithm with local search (PGA-ACS) algorithm for multi-objective FSP with an objective of minimizing the makespan and total flowtime. Loukil et al. (2005) proposed multi-objective simulated annealing algorithm to tackle the multi-objective production scheduling problems.

Pareto dominance: Among a set of schedules P, a schedule $x^1 \in P$ is said to *dominate* the other schedule $x^2 \in P$, denoted as $\left(x^1 \phi x^2\right)$, if both the following conditions are true.

(i) The schedule $x^1 \in P$ is no worse than $x^2 \in P$ in all objectives.

(ii) The schedule $x^1 \in P$ is strictly better than $x^2 \in P$ in at least one objective.

When both the conditions are satisfied, x^2 is called as a dominated schedule and x^1 a non-dominated schedule. If any of the above condition is violated, the schedule x^1 does not dominate the schedule x^2. Among a set of schedules P, the non-dominated set P' are those that are not dominated by any member of the set (Deb, 2003).

Non-dominated front: The set of all non-dominated schedules.

Pareto optimal set: When the set P is the entire search space X, the resulting non-dominated set is called the Pareto optimal set.

The primary objective is to find a set of non-dominated fronts for the FSPs with the consideration of performance measures.

5.2 Proposed Multi-objective Discrete PSO Algorithm

The discrete PSO algorithm proposed for single objective FSP has been suitably modified to generate non-dominated solution set considering three performance measures simultaneously. Before presenting the proposed algorithm, the non-dominated sorting procedure, Pareto search procedure and the parameters considered are discussed below.

***Non- Domination Sorting*:** Non-domination measures are used to find non-dominated set of solutions. The following procedure is used to generate non-dominated particle or solution set from the population of particles. Consider a swarm consisting of N solutions (particles).

Step 0: Begin with $i = 1; j = i + 1$, and repeat steps 1 and 2.

Step 1: Compare solutions x^i and x^j for domination using the two conditions mentioned.

Step 2: If x^j is dominated by x^i, mark x^j as 'dominated,' increment j, and go to step 1. Otherwise mark x^i as dominated, increment i, set $j = i + 1$ and go to step 1.

All solutions that are not marked 'dominated' forms a non-dominated solution set and these are stored separately in a memory called archive.

```
Initialize the parameters
Generate the swarm and velocity
t = 0: // iteration counter
Evaluate all the particles
Perform non-dominated sorting to identify G_b^t
Open Archive to store G_b^t

Do    {
        Update position;
        t = t + 1
        Evaluate
        Do non-dominated sorting to identify G_b^t
        Archive update
        Update velocity
}     while (t < t_max) :  t_max = 100;
Output G_b^t
```

Figure 6. Iterative search loop of the multi-objective discrete PSO algorithm

***Pareto Search*:** In case of a single objective scheduling optimization, an optimal solution forms the Global best (G_b^t). Under multi-objective scheduling, with multiple objectives, G_b^t consist of a set of non-dominated solutions. Once the swarm is initialized, $G_b\,(t = o)$ is obtained after non-dominated sorting of the particles. During the subsequent iterations, position and velocity update of the particles are carried out using local best and global best. It is to be noted that one solution is randomly chosen from the archive as Global best set. During every iteration, non-dominated solution set is updated. This non-dominated solution set is added with the Archive and the combined set is sorted for non-dominance. Dominated solutions within the combined set are removed and the remaining non-dominated solutions forms $G_b\,(t = 1)$. This procedure is repeated to guide the non-dominated search process towards the Pareto region. Initially, a set of particles are generated randomly and evaluated. Then the non-dominated sorting of particles is done. Within the swarm, the non-dominated solution set i.e. G_b^t is identified and they are stored in an archive. Then the positions and velocities of the particles are updated iteratively. These current sets of non-dominated solutions are combined with the archive solutions. Non-dominated sorting of archive is done to identify the archive survival members. This process is called Archive update. During this, all dominated members of the combined set are removed. This procedure is repeated to guide the non-dominated search process towards generating a solution front close to the Pareto region. After the termination criterion is met, the solution set stored in the archive forms the result. The iterative improvement process of multi-objective PSO algorithm is presented in Figure 6.

5.3 Performance of Multi-objective Discrete PSO Algorithm

In this section, the performance measures namely minimization of makespan, total flowtime and completion time variance are considered simultaneously. It is to be noted that PSO algorithm has been very rarely studied by researchers for solving FSPs with multi-objective requirements.

***Parameter Selection*:** Using the proposed algorithm, experiments are conducted to redesign the algorithm with appropriate parameter settings. Parameters were identified by trial and error approach for the better performance. The swarm size is taken as 80. The values of acceleration constants are fixed by trial and error as $C_1 = 1; C_2 = 2$ and $C_3 = 2$. The values of velocity coefficients U_1, U_2 and U_3 are generated randomly between 0 and 1. Termination criterion is taken as 100 iterations. The benchmark instances of Taillard (1993) form a set of 120 problems of various sizes, having 20, 50, 100, 200 and 500 jobs and 5, 10 or 20 machines have been taken and solved. When the iterative search process is continued beyond 100 iterations, solution quality is expected to improve further and the non-dominated front will converge towards the Pareto front. Some samples of non-dominated solution sets obtained during 1st, 50th and 100th iterations of selected benchmark FSPs are presented in Table 6. to Table 10.

1st Iteration			50th Iteration			100th Iteration		
C_{max}	F_Σ	V_T	C_{max}	F_Σ	V_T	C_{max}	F_Σ	V_T
2372	37335	143185.03	2418	37282	163238.72	2380	37749	121770.54
2385	37379	134831.95	2450	37645	139131.23	2395	37465	130522.04
2410	36900	148013.59	2451	38181	137003.64	2458	37187	210477.03
2412	37674	129799.71	2495	36838	186985.58	2465	37341	187537.33
2412	36970	138733.95	2518	39668	127576.64	2488	36988	148466.05
2414	36786	157977.52	2544	36566	258462.42	2493	36787	244247.03
2425	36842	155318.20	2550	36352	180610.66	2518	36639	213526.66
2432	36071	225477.25	2633	37206	175815.11	2545	36177	189031.61
2437	36855	150071.23						
2448	37604	125025.85						
2451	36764	158552.28						
2451	36600	172676.80						
2464	37521	134748.27						
2468	37875	124452.44						
2480	39012	119837.64						
2491	36170	154730.75						
2523	38802	123177.59						

Table 6. Non-dominated fronts obtained for 20 x 20 FSP (Problem ta025 of Taillard,1993)

1st Iteration			50th Iteration			100th Iteration		
C_{max}	F_Σ	V_T	C_{max}	F_Σ	V_T	C_{max}	F_Σ	V_T
3840	127915	674615.63	4168	138549	801359.25	4192	143728	688058.06
3923	132364	655699.19	4170	139913	794893.25	4218	142073	835741.13
3979	130656	669600.50	4181	140250	769993.81	4226	136757	870648.81
3979	132435	633633.38	4188	138913	756248.50	4241	140962	788543.19
3982	132026	666358.94	4243	137007	882535.81	4245	138443	845496.63
4018	136354	604771.06	4254	141017	750998.31	4266	137938	828836.88
4023	132426	646723.94	4284	136183	929310.25	4298	137356	866164.31
4034	135781	631409.19	4290	137714	833303.44	4324	143038	776172.63
4058	131370	652795.69	4295	135927	845500.88	4329	143586	760850.94
4081	137607	586079.44	4319	142649	731565.19	4334	141675	780154.75
4084	136148	601373.06	4320	140119	747898.00	4343	136398	868004.75

Table 7. Non-dominated fronts obtained for 50 x 20 FSP (Problem ta055 of Taillard,1993)

1st Iteration			50th Iteration			100th Iteration		
C_{max}	F_Σ	V_T	C_{max}	F_Σ	V_T	C_{max}	F_Σ	V_T
6719	414626	2332780.00	7079	442243	2714971.00	6977	429237	2643600.00
6736	407661	2339133.25	7122	431015	2619110.50	7187	429079	2992237.75
6754	407217	2426269.50	7125	430238	2888681.25	7222	423655	3181877.50
6759	414920	2322475.00	7279	427670	3036344.25	7266	427705	3032460.25
6772	421227	2319961.50	7307	426737	3014873.00	7287	426588	3061585.25
6776	420444	2215965.00						
6780	406735	2308902.00						
6785	417764	2299484.50						
6804	417373	2165440.25						
6934	402802	2477583.00						

Table 8. Non-dominated fronts obtained for 100 x 20 FSP (Problem ta085 of Taillard,1993)

1st Iteration			50th Iteration			100th Iteration		
C_{max}	F_Σ	V_T	C_{max}	F_Σ	V_T	C_{max}	F_Σ	V_T
11883	1341992	8681969.00	12169	1370395	8968974.00	12213	1382492	9226709.00
11922	1378165	8301979.00				12246	1418388	8839896.00
11935	1361242	8654574.00				12304	1390924	9191086.00
11938	1365058	8581394.00				12361	1380781	9530417.00
11964	1363602	8492216.00				12445	1379004	9589141.00
11995	1355612	8551758.00						
12020	1371423	8237680.50						
12051	1369441	8470111.00						
12115	1354810	8405068.00						

Table 9. Non-dominated fronts obtained for 200 x 20 FSP (Problem ta105 of Taillard,1993)

1st Iteration			50th Iteration			100th Iteration		
C_{max}	F_Σ	V_T	C_{max}	F_Σ	V_T	C_{max}	F_Σ	V_T
27361	7380460	53524660.00	27802	7498389	54440864.00	27612	7421436	53180528.00
27417	7405289	51892856.00	27811	7402468	53268448.00	27765	7458248	53042776.00
27448	7419382	51504108.00	27999	7543786	53059836.00	27870	7440681	53140668.00
27465	7394286	52016468.00	28091	7529455	52754652.00	27891	7374759	53306856.00
27534	7392887	51930096.00						
27593	7458730	51066888.00						
27603	7373445	51681608.00						
27638	7439401	51390116.00						
27680	7445450	51262332.00						
27700	7418177	51122680.00						
27729	7492150	51039416.00						

Table 10.Non-dominated fronts obtained for 500 x 20 FSP (Problem ta115 of Taillard,1993)

Normalized values of the performance measures are plotted for better visualization. Some samples of non-dominated front obtained during 1st, 50th and 100th iterations of selected benchmark FSPs are presented in Fig. 7. to Fig. 11.

6. Conclusion

Literature survey indicates that very few authors have studied the applications of multi-objective scheduling in flowshop scheduling using particle swarm optimization algorithm is scarce. This Chapter presents a discrete PSO algorithm to solve FSPs. This work has been conducted in two phases. In the first phase, a discrete PSO is proposed to solve the single-objective FSPs. In the second phase, a multi-objective discrete PSO algorithm is proposed to solve the FSPs with three objectives. The performance of the proposed single-objective discrete PSO is tested by solving a large set of benchmark FSPs. The quality measure namely "Average Relative Percent Deviation" ($\overline{\text{RPD}}$) is used to compare the solution quality obtained with the results available in the literature. It shows that the proposed discrete PSO algorithm performs better in terms of quality of results. Using the proposed algorithm, experiments are conducted to redesign the algorithm with appropriate parameter settings. The $\overline{\text{RPD}}$ for each set of instances are also shown in an efficient way. The parameters selected for solving the problems are holds good. The proposed multi-objective discrete PSO algorithm performs better in terms of yielding more number of non-dominated solutions close to Pareto front during the search. It is seen that, when the number of iterations is more, the non-dominated solution set generated is close to the Pareto front.

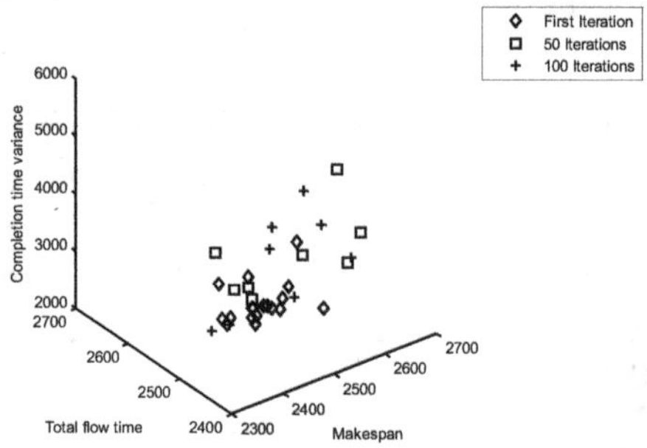

Figure 7. Non-dominated solution set obtained for 20 x 20 FSP (Problem ta025 of Taillard,1993)

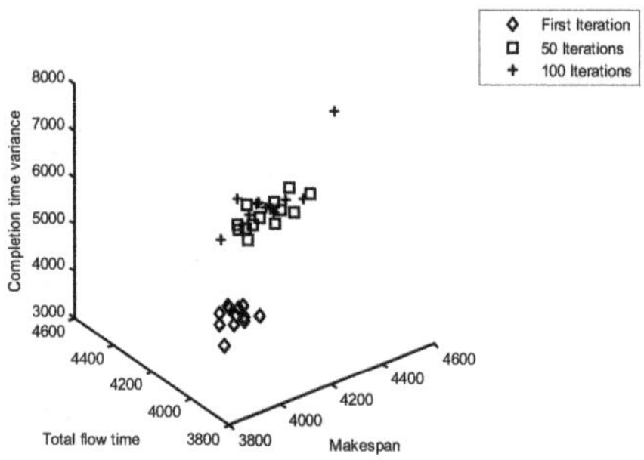

Figure 8. Non-dominated solution set obtained for 50 x 20 FSP (Problem ta055 of Taillard,1993)

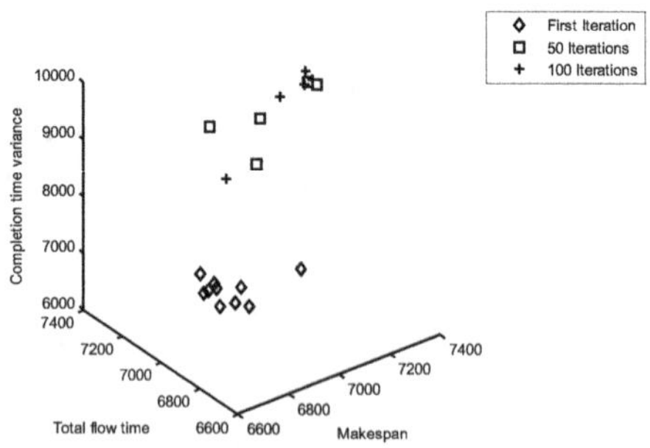

Figure 9. Non-dominated solution set obtained for 100x20 FSP (Problem ta085 of Taillard,1993)

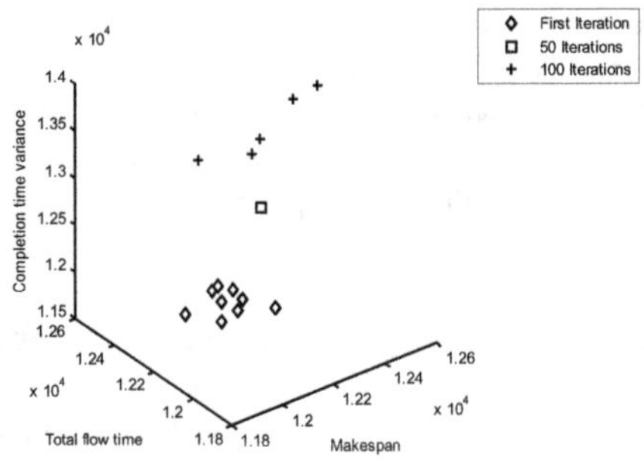

Figure 10.Non-dominated solution set obtained for 200x20 FSP (Problem ta105 of Taillard,1993)

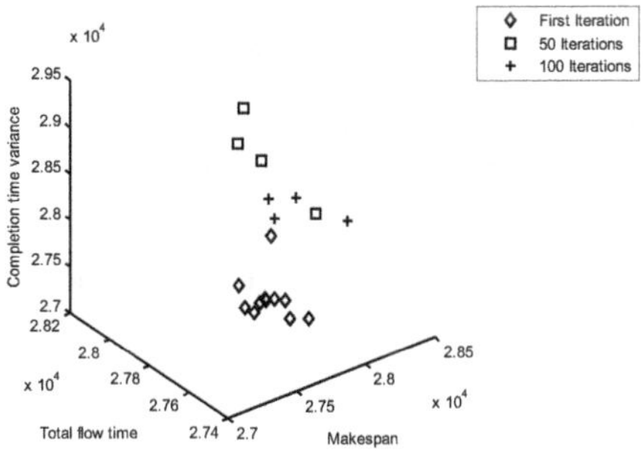

Figure 11.Non-dominated solution set obtained for 500x20 FSP (Problem ta115 of Taillard,1993)

7. References

Abido, M.A. (2002). Optimal power flow using particle swarm optimization. *Electrical Power and Energy Systems*, Vol.24, 563-571

Bagchi, T.P. (1999). *Multi-objective scheduling by Genetic Algorithms*, Kluwer Academic Publishers, Boston, Massachusetts

Baker, K.R. (1974). *Introduction to Sequencing and Scheduling*, John Wiley & Sons, New York

Brucker, P. (1998). *Scheduling Algorithms*, Springer-Verlag, Berlin

Campbell, H.G.; Dudek, R.A. & Smith, M.L. (1970). A heuristic algorithm for the n job, m machine sequencing problem, *Management Science*, Vol.16, No: 10, B630-B637

Chan, F.T.S.; Wong, T.C. & Chan, L.Y. (2005). A genetic algorithm based approach to machine assignment problem, *International Journal of Production Research*, Vol.43, No: 12, 2451-2472

Chandrasekaran, S.; Ponnambalam, S.G.; Suresh, R.K. & Vijayakumar N. (2006). An Application of Particle Swarm Optimization Algorithm to Permutation Flowshop Scheduling Problems to Minimize Makespan, Total Flowtime and Completion Time Variance, *Proceedings of the IEEE International Conference on Automation Science and Engineering, 2006 (CASE '06.)*, pp-513-518, ISBN: 1-4244-0311-1, Shanghai, China,

Chandrasekharan Rajendran. & Hans Ziegler. (2005). Two Ant-colony algorithms for minimizing total flowtime in permutation flowshops, *Computers & Industrial engineering*, Vol.48, 789-797

Christian Blum. & Andrea Roli. (2003). Metaheuristics in Combinatorial Optimization: Overview and Conceptual Comparison. *ACM Computing Surveys*, Vol. 35, No. 3, 268-309

Dannenbring, D.G. (1977). An evaluation of flowshop sequencing heuristics, *Management Science*, Vol.23, No: 11, 1174-1182

Dudek, R.A.; Panwalkar, S.S. & Smith, M.L. (1992). The lessons of flowshop scheduling research, *Operations Research*, Vol.40, No: 1, 7-13

Eberhart, R.C. & Hu, X. (1999). Human tremor analysis using particle swarm optimization. *Proceedings of the Congress on Evolutionary Computation*, pp-1927-1930, IEEE Service Center, Washington, DC, Piscataway, NJ

Eberhart, R.C. & Kennedy J. (1995). A New Optimizer Using Particles Swarm Theory, *Proceedings of the Sixth International Symposium on Micro Machine and Human Science*, pp-39-43, IEEE Service Center, Nagoya, Japan

Eberhart, R.C. & Shi, Y. (2001). Particle swarm optimization: developments, applications and resources, *Proceedings of IEEE Congress on Evolutionary Computation 2001*, Seoul, Korea

Faith Tasgetiren, S.; Mehmet Sevkli.; Yen-Chia Liang. & Gunes Gencyilmaz. (2004). Particle swarm optimization algorithm for single machine total weighted tardiness problem, *IEEE Transaction on Power and Energy Systems*, 1412-1419

Framinan, J.M. & Leisten, R. (2003). An efficient constructive heuristic for flowtime minimization in permutation flowshops, *Omega*, Vol.31, 311-317

French, S. (1982) *Sequencing and Scheduling: An introduction to the mathematics of the jobshop*, Ellis Horword Limited, Chichester, England

Gowrishankar, K.; Rajendran, C. & Srinivasan, G. (2001). Flowshop scheduling algorithms for minimizing the completion time variance and the sum of squares of completion time deviation from the common due date, *European Journal of Operational Research*, vol.132, No: 31, 643-665

Ignall, E. & Scharge, L. (1965). Application of the branch and bound technique to some flowshop-scheduling problems, *Operations Research*, Vol.13, 400-412

Ishibuchi, H.; Yoshida, T. & Murata, T. (2003). Balance between genetic search and local search in memetic algorithms for multi-objective permutation flowshop scheduling, *IEEE Transaction on Evolutionary Computation*, Vol.7 No.2, 204-223

Johnson, S.M. (1954). Optimal two-stage and three-stage production schedules with setup times included, *Naval Research Logistics Quarterly*, Vol.1 61-68

Kalyanmoy Deb. (2003). *Multi-objective Optimization Using Evolutionary Algorithms*, John Wiley & Sons, First Edition.

Kennedy, J. & Eberhart, R. (1995). Particle swarm optimization, *Proceedings of IEEE International Conference on Neural Networks-IV*, pp-1942-1948, Piscataway, NJ: IEEE service center, Perth, Australia

Kennedy, J.; Eberhart, R. & Shi, Y. (2001). *Swarm Intelligence*, Morgan Kaufmann, San Mateo,CA,USA

Liu, J. & Reeves, C.R. (2001). Constructive and composite heuristic solutions to the $P//\sum C_i$ scheduling problem, *European Journal of Operational Research.*, Vol.132, 439-452

Lixin Tang. & Jiyin Liu. (2002). A modified genetic algorithm for the flowshop sequencing problem to minimize mean flowtime, *Journal of Intelligent Manufacturing*, Vol.13, 61-67

Loukil, T.; Teghem, J. & Tuyttens, D. (2005). Solving multi-objective production scheduling problems using metaheuristics, *European Jour. of Operational Research*, Vol.161, 42-61

Merkle, D. & Middendorf, M. (2000). An ant algorithm with new pheromone evaluation rule for total tardiness problems, *Proceedings of the Evolutionary Workshops 2000*, pp-287-296, vol.1803, Lecture Notes in Computer Science, Springer

Moccellin, J.V. & Nagano, M.S. (1998). Evaluating the performance of tabu search procedures for flowshop sequencing, *Journal of the Operational Research Society*, Vol.49, 1296-1302

Nawaz, M.; Enscore Jr, E.E. & Ham, I. (1983). A Heuristic algorithm for the m-machine, n-job scquencing problem, *Omega*, Vol.11, 91-98

Ogbu, F.A. & Smith, D.K. (1990). The application of the simulated annealing algorithm to the solution of the $n/m/C_{max}$ flowshop problem, *Computers and Operations Research*, Vol.17, No: 3, 243-253

Osman, I.H. & Laporte, G. (1996). Metaheuristics: A bibliography. *Operations Research*, Vol.63, 513–623

Palmer, D. (1965). Sequencing jobs through a multi-stage process in the minimum total time- a quick method of obtaining a near optimum, *Opn. Research*, Vol.16, No: 1, 101-107

Pasupathy, T.; Chandrasekharan Rajendran. & Suresh, R.K. (2006). A multi-objective genetic algorithm for scheduling in flowshops to minimize makespan and total flowtime, *International Journal of Advanced Manufacturing Technology*, Springer-Verlag London Ltd, Vol.27, 804-815

Pinedo, M. (2002). *Scheduling: Theory, Algorithms and Systems*, Second edition,. Prentice-Hall, Englewood Cliffs, New Jersey

Prabhaharan, G.; Shahul Hamid Khan, B.; Asokan, P. & Thiyagu M. (2005). A Particle swarm optimization algorithm for permutation flowshop scheduling with regular and non-regular measures, *International Journal of Applied Management and Technology*, Vol.3, No: 1, 171-182

Rajendran, C., (1993). Heuristic algorithm for scheduling in a flowshop to minimize total flowtime, *International Journal of Production Economics*, Vol.29, 65-73

Rameshkumar, K.; Suresh, R.K. & Mohanasundaram, K.M. (2005). Discrete particle swarm optimization (DPSO) algorithm for permutation flowshop scheduling to minimize makespan, *Lecture Notes in Comp. Science*, Springer Verlag-GMBH.0302-9743. Vol.3612

Rinnooy Kan, A.H.G. (1976). *Machine Scheduling Problems: Classification, Complexity and Computations*, Nijhoff, The Hague

Ruben Ruiz. & Concepcion Maroto. (2005). A comprehensive review and evaluation of permutation flowshop heuristics, *European Journal of operational Research*, Vol.165, 479-494

Ruben Ruiz.; Concepcion Maroto. & Javier Alcaraz. (2004). Two new robust genetic algorithms for the flowshop scheduling problem, *OMEGA*, 2-16

Sensarma, P. S. ; Rahmani, M. & Carvalho, A. (2002). A comprehensive method for optimal expansion planning using particle swarm optimization, *IEEE Power Engineering Society Winter Meeting*, Vol. 2, .1317-1322

Sridhar, J. & Rajendran, C. (1996). Scheduling in flowshop and cellular manufacturing system with multiple objectives - A genetic algorithmic approach, *Production Planning and Control*, Vol.74, 374-382

Stuetzle, T. (1998). An ant approach for the flowshop problem, *Proceedings of the 6th European Congress on Intelligent Techniques and Soft Computing* (EUFIT '98), pp-1560-1564, Vol.3, Verlag Mainz, Aachen, Germany

Suresh, R.K, & Mohanasundaram, K.M, (2004). Pareto archived simulated annealing for permutation flowshop scheduling with multiple objectives, *Proceedings of the IEEE Conference on Cybermatics and Intelligent Systems*, pp-1-3, Singapore

Taillard, E. (1993). Benchmarks for basic scheduling problem, *European Journal of Operational Research*, Vol.64, 278-285

Tandon, V. (2000). Closing the gap between CAD/CAM and optimized CNC end milling. *Master's thesis*, Purdue School of Engineering and Technology, Indiana University ,Purdue University, Indianapolis.

Yoshida, H.; Kawata, K.; Fukuyama, Y. & Nakanishi, Y. (1999). A particle swarm optimization for reactive power and voltage control considering voltage stability. *Proceedings of the International Conference on Intelligent System Application to Power Systems*, pp-117-121, Rio de Janeiro, Brazil

Yuhui Shi. (2004). Particle Swarm Optimization, *IEEE Neural Networks Society*, 8-13

Yuvraj Gajpal. & Chandrasekharan Rajendran. (2006). An ant-colony optimization algorithm for minimizing the completion time variance of jobs in flowshops, *International Journal of Production Economics*, Vol. 101, No: 2, 259-272

14

A Particle Swarm Optimization Technique used for the Improvement of Analogue Circuit Performances

Mourad Fakhfakh[1], Yann Cooren[2], Mourad Loulou[1] and Patrick Siarry[2]
[1]University of Sfax, [2]University of Paris 12
[1]Tunisia, [2]France

1. Introduction

The importance of the analogue part in integrated electronic systems cannot be overstressed. Despite its eminence, and unlike the digital design, the analogue design has not so far been automated to a great extent, mainly due to its towering complexity (Dastidar et al., 2005). Analogue sizing is a very complicated, iterative and boring process whose automation is attracting great attention (Medeiro et al., 1994). The analogue design and sizing process remains characterized by a mixture of experience and intuition of skilled designers (Tlelo-Cuautle & Duarte-Villaseñor, 2008). As a matter of fact, optimal design of analogue components is over and over again a bottleneck in the design flow.

Optimizing the sizes of the analogue components automatically is an important issue towards ability of rapidly designing true high performance circuits (Toumazou & Lidgey, 1993; Conn et al., 1996).

Common approaches are generally either fixed topology ones or/and statistical-based techniques. They generally start with finding a "good" DC quiescent point, which is provided by the skilled analogue designer. After that a simulation-based tuning procedure takes place. However these statistic-based approaches are time consuming and do not guarantee the convergence towards the global optimum solution (Talbi, 2002).

Some mathematical heuristics were also used, such as Local Search (Aarts & Lenstra, 2003), Simulated Annealing (Kirkpatrick et al., 1983; Siarry(a) et al., 1997), Tabu Search (Glover, 1989; Glover, 1990), Genetic Algorithms (Grimbleby, 2000; Dréo et al., 2006), etc. However these techniques do not offer general solution strategies that can be applied to problem formulations where different types of variables, objectives and constraint functions are used. In addition, their efficiency is also highly dependent on the algorithm parameters, the dimension of the solution space, the convexity of the solution space, and the number of variables.

Actually, most of the circuit design optimization problems simultaneously require different types of variables, objective and constraint functions in their formulation. Hence, the abovementioned optimization procedures are generally not adequate or not flexible enough.

In order to overcome these drawbacks, a new set of nature inspired heuristic optimization algorithms were proposed. The thought process behind these algorithms is inspired from the collective behaviour of decentralized, self-organized systems. It is known as Swarm Intelligence (SI) (Bonabeau et al. 1999). SI systems are typically made up of a population of simple agents (or ''particles'') interacting locally with each other and with their environment. These particles obey to very simple rules, and although there is no centralized control structure dictating how each particle should behave, local interactions between them lead to the emergence of complex global behaviour. Most famous such SIs are Ant Colony Optimization (ACO) (Dorigo et al., 1999), Stochastic Diffusion Search (SDS) (Bishop, 1989) and Particle Swarm Optimization (PSO) (Kennedy & Eberhart, 1995; Clerc, 2006).

PSO, in its current form, has been in existence for almost a decade, which is a relatively short period when compared to some of the well known natural computing paradigms, such as evolutionary computation. PSO has gained widespread demand amongst researchers and has been shown to offer good performance in an assortment of application domains (Banks et al., 2007).

In this chapter, we focus on the use of PSO technique for the optimal design of analogue circuits. The practical applicability and suitability of PSO to optimize performances of such multi-objective problems are highlighted. An example of optimizing performances of a second generation MOS current conveyor (CCII) is presented. The used PSO algorithm is detailed and Spice simulation results, performed using the 'optimal' sizing of transistors forming the CCII and bias current, are presented. Reached performances are discussed and compared to others presented in some published works, but obtained using classical approaches.

2. The Sizing Problem

The process of designing an analogue circuit mainly consists of the following steps (Medeiro et al., 1994) :

- the topology choice: a suitable schematic has to be selected,
- the sizing task: the chosen schematic must be dimensioned to comply with the required specifications,
- The generation of the layout.

Among these major steps, we focus on the second one, i.e. the optimal sizing of analogue circuits.

Actually, analogue sizing is a constructive procedure that aims at mapping the circuit specifications (objectives and constraints on performances) into the design parameter values. In other words, the performance metrics of the circuit, such as gain, noise figure, input impedance, occupied area, etc. have to be formulated in terms of the design variables (Tulunay & Balkir, 2004).

In a generic circuit, the optimization problem consists of finding optimal values of the design parameters. These variables form a vector $\vec{X}^T = \{x_1, x_2, \ \ , x_N\}$ belonging to an N-dimensional design space. This set includes transistor geometric dimensions and passive component values, if any. Hence, performances and objectives involved in the design objectives are expressed as functions of X.

These performances may belong to the set of constraints ($\vec{g}(\vec{X})$) and/or to the set of objectives ($\vec{f}(\vec{X})$). Thus, a general optimization problem can be formulated as follows:

$$\begin{aligned} &\textit{minimize} &:\quad & f_i(\vec{X}),\ i\in[1,k] \\ &\textit{such that:} &:\quad & g_j(\vec{X})\le 0,\ j\in[1,l] \\ &&& h_m(\vec{X})\le 0,\ m\in[1,p] \\ &&& x_{Li}\le x_i\le x_{Ui},\ i\in[1,N] \end{aligned} \tag{1}$$

k, l and p denote the numbers of objectives, inequality constraints and equality constraints, respectively. $\vec{x}_L$ and $\vec{x}_U$ are lower and upper boundaries vectors of the parameters.

The goal of optimization is usually to minimize an objective function; the problem for maximizing $\vec{f}(\vec{x})$ can be transformed into minimizing $-\vec{f}(\vec{x})$. This goal is reached when the variables are located in the set of optimal solutions.

For instance, a basic two-stage operational amplifier has around 10 parameters, which include the widths and lengths of all transistors values which have to be set. The goal is to achieve around 10 specifications, such as gain, bandwidth, noise, offset, settling time, slew rate, consumed power, occupied area, CMRR (common-mode rejection ratio) and PSRR (power supply rejection ratio). Besides, a set of DC equations and constraints, such as transistors' saturation conditions, have to be satisfied (Gray & Meyer, 1982).

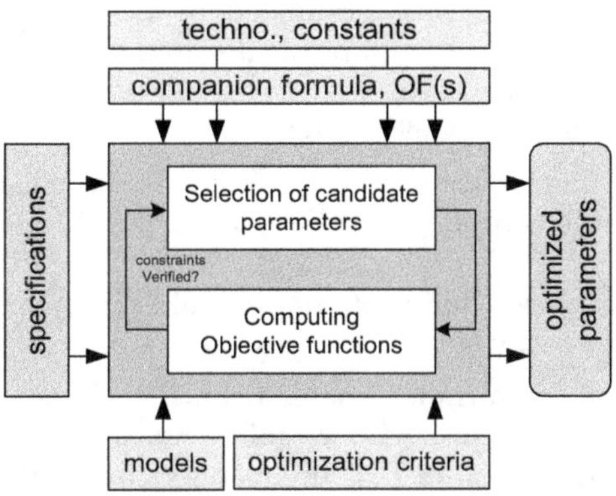

Figure 1. Pictorial view of a design optimization approach

The pictorial flow diagram depicted in Fig. 1 summarizes main steps of the sizing approach.

As it was introduced in section 1, there exist many papers and books dealing with mathematic optimization methods and studying in particular their convergence properties (see for example (Talbi, 2002; Dréo et al., 2006; Siarry(b) et al., 2007)).

These optimizing methods can be classified into two categories: deterministic methods and stochastic methods, known as heuristics.

Deterministic methods, such as Simplex (Nelder & Mead, 1965), Branch and Bound (Doig, 1960), Goal Programming (Scniederjans, 1995), Dynamic Programming (Bellman, 2003)... are effective only for small size problems. They are not efficient when dealing with NP-hard and multi-criteria problems. In addition, it has been proven that these optimization techniques impose several limitations due to their inherent solution mechanisms and their tight dependence on the algorithm parameters. Besides they rely on the type of objective, the type of constraint functions, the number of variables and the size and the structure of the solution space. Moreover they do not offer general solution strategies.

Most of the optimization problems require different types of variables, objective and constraint functions simultaneously in their formulation. Therefore, classic optimization procedures are generally not adequate.

Heuristics are necessary to solve big size problems and/or with many criteria (Basseur et al., 2006). They can be 'easily' modified and adapted to suit specific problem requirements. Even though they don't guarantee to find in an exact way the optimal solution(s), they give 'good' approximation of it (them) within an acceptable computing time (Chan & Tiwari, 2007). Heuristics can be divided into two classes: on the one hand, there are algorithms which are specific to a given problem and, on the other hand, there are generic algorithms, i.e. metaheuristics. Metaheuristics are classified into two categories: local search techniques, such as Simulated Annealing, Tabu Search … and global search ones, like Evolutionary techniques, Swarm Intelligence techniques …

ACO and PSO are swarm intelligence techniques. They are inspired from nature and were proposed by researchers to overcome drawbacks of the aforementioned methods. In the following, we focus on the use of PSO technique for the optimal design of analogue circuits.

3. Overview of Particle Swarm Optimization

The particle swarm optimization was formulated by (Kennedy & Eberhart, 1995). The cogitated process behind the PSO algorithm was inspired by the optimal swarm behaviour of animals such, as birds, fishes and bees.

PSO technique encompasses three main features:

- It is a SI technique; it mimics some animal's problem solution abilities,
- It is based on a simple concept. Hence, the algorithm is neither time consumer nor memory absorber,
- It was originally developed for continuous nonlinear optimization problems. As a matter of fact, it can be easily expanded to discrete problems.

PSO is a stochastic global optimization method. Like in Genetic Algorithms (GA), PSO exploits a population of potential candidate solutions to investigate the feasible search space. However, in contrast to GA, in PSO no operators inspired by natural evolution are applied to extract a new generation of feasible solutions. As a substitute of mutation, PSO relies on the exchange of information between individuals (particles) of the population (swarm).

During the search for the promising regions of the landscape, and in order to tune its trajectory, each particle adjusts its velocity and its position according to its own experience, as well as the experience of the members of its social neighbourhood. Actually, each particle remembers its best position, and is informed of the best position reached by the swarm, in the global version of the algorithm, or by the particle's neighbourhood, in the local version of the algorithm. Thus, during the search process, a global sharing of information takes place and each particle's experience is thus enriched thanks to its discoveries and those of all the other particles. Fig. 2 illustrates this principle.

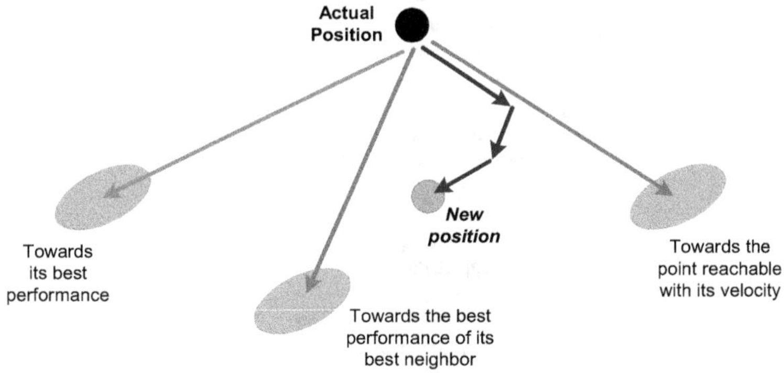

Figure 2. Principle of the movement of a particle

In an N-dimensional search space, the position and the velocity of the ith particle can be represented as $X_i = [x_{i,1}, x_{i,2}, \ldots, x_{i,N}]$ and $V_i = [v_{i,1}, v_{i,2}, \ldots, v_{i,N}]$ respectively. Each particle has its own best location $P_i = [p_{i,1}, p_{i,2}, \ldots, p_{i,N}]$, which corresponds to the best location reached by the ith particle at time *t*. The global best location is named $g = [g_1, g_2, \ldots, g_N]$, which represents the best location reached by the entire swarm. From time *t* to time *t+1*, each velocity is updated using the following equation:

$$v_{i,j}(t+1) = \overbrace{w\, v_{i,j}(t)}^{inertia} + \overbrace{c_1\, r_1 (p_{i,j} - v_{i,j}(t))}^{Personal\ Influence} + \overbrace{c_2\, r_2 (g_i - v_{i,j}(t))}^{Social\ Influence} \tag{2}$$

where w is a constant known as inertia factor, it controls the impact of the previous velocity on the current one, so it ensures the diversity of the swarm, which is the main means to avoid the stagnation of particles at local optima. c_1 and c_2 are constants called acceleration coefficients; c_1 controls the attitude of the particle of searching around its best location and c_2 controls the influence of the swarm on the particle's behaviour. r_1 and r_2 are two independent random numbers uniformly distributed in [0,1].

The computation of the position at time *t+1* is derived from expression (2) using:

$$x_{i,j}(t+1) = x_{i,j}(t) + v_{i,j}(t+1) \tag{3}$$

It is important to put the stress on the fact that the PSO algorithm can be used for both mono-objective and multi-objective optimization problems.

The driving idea behind the multi-objective version of PSO algorithm (MO-PSO) consists of the use of an archive, in which each particle deposits its 'flight' experience at each running cycle. The aim of the archive is to store all the non-dominated solutions found during the optimization process. At the end of the execution, all the positions stored in the archive give us an approximation of the theoretical Pareto Front. Fig. 3 illustrates the flowchart of the MO-PSO algorithm. Two points are to be highlighted: the first one is that in order to avoid excessive growing of the storing memory, its size is fixed according to a crowding rule (Cooren et al., 2007). The second point is that computed optimal solutions' inaccuracy crawls in due to the inaccuracy of the formulated equations.

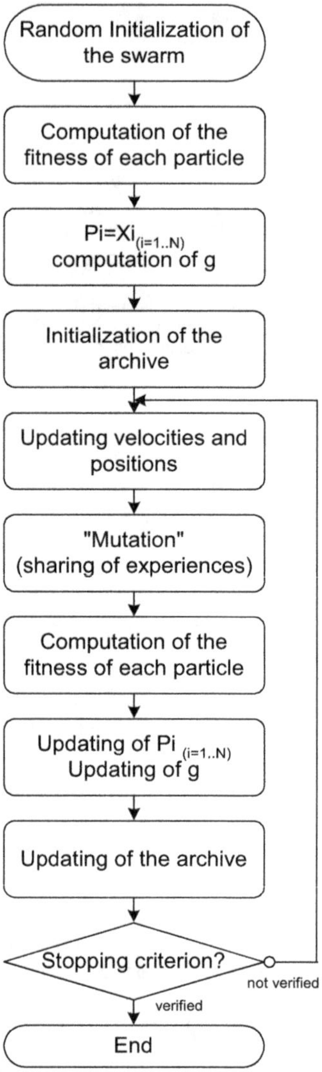

Figure 3. Flowchart of a MO-PSO

In the following section we give an application example dealing with optimizing performances of an analogue circuit, i.e. optimizing the sizing of a MOS inverted current conveyor in order to maximize/minimize performance functions, while satisfying imposed and inherent constraints. The problem consists of generating the trade off surface (Pareto front[1]) linking two conflicting performances of the CCII, namely the high cut-off current frequency and the parasitic X-port input resistance.

[1] Definition of Pareto optimality is given in Appendix.

4. An Application Example

The problem consists of optimizing performances of a second generation current conveyor (CCII) (Sedra & Smith, 1970) regarding to its main influencing performances. The aim consists of maximizing the conveyor high current cut-off frequency and minimizing its parasitic X-port resistance (Cooren et al., 2007).
In the VLSI realm, circuits are classified according to their operation modes: voltage mode circuits or current mode circuits. Voltage mode circuits suffer from low bandwidths arising due to the stray and circuit capacitances and are not suitable for high frequency applications (Rajput & Jamuar, 2007).
In contrary, current mode circuits enable the design of circuits that can operate over wide dynamic ranges. Among the set of current mode circuits, the current conveyor (CC) (Smith & Sedra, 1968; Sedra & Smith, 1970) is the most popular one.
The Current Conveyor (CC) is a three (or more) terminal active block. Its conventional representation is shown in Fig. 4a. Fig. 4b shows the equivalent nullator/norator representation (Schmid, 2000) which reproduces the ideal behaviour of the CC. Fig. 4.c shows a CCII with its parasitic components (Ferry et al. 2002).

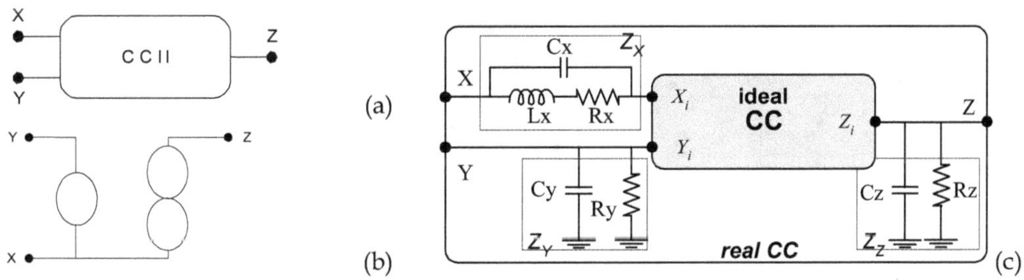

Figure 4. (a) General representation of current conveyor, (b) the nullor equivalency: ideal CC, (c) parasitic components: real CC

Relations between voltage and current terminals are given by the following matrix relation (Toumazou & Lidgey, 1993):

$$\begin{bmatrix} I_Y \\ V_X \\ I_Z \end{bmatrix} = \begin{bmatrix} \frac{1}{(C_Y // R_Y)} & \alpha & 0 \\ \gamma & R_X & 0 \\ 0 & \beta & \frac{1}{(C_Z // R_Z)} \end{bmatrix} \begin{bmatrix} V_Y \\ I_X \\ V_Z \end{bmatrix} \tag{4}$$

For the above matrix representation, α specifies the kind of the conveyor. Indeed, for α =1, the circuit is considered as a first generation current conveyor (CCI). Whereas when α =0, it is called a second generation current conveyor (CCII). β characterizes the current transfer from X to Z ports. For β=+1, the circuit is classified as a positive transfer conveyor. It is considered as a negative transfer one when β=-1. γ=±1: When γ=-1 the CC is said an inverted CC, and a direct CC, otherwise.

Accordingly, the CCII ensures two functionalities between its terminals:

- A Current follower/Current mirror between terminals X and Z.
- A Voltage follower/Voltage mirror between terminals X and Y.

In order to get ideal transfers, CCII are commonly characterized by low impedance on terminal X and high impedance on terminals Y and Z.

In this application we deal with optimizing performances of an inverted positive second generation current conveyor (CCII+) (Sedra & Smith, 1970; Cooren et al., 2007) regarding to its main influencing performances. The aim consists of determining the optimal Pareto circuit's variables, i.e. widths and lengths of each MOS transistor, and the bias current I_0, that maximizes the conveyor high current cut-off frequency and minimizes its parasitic X-port resistance (R_X) (Bensalem et al., 2006; Fakhfakh et al. 2007). Fig. 5 illustrates the CCII+'s MOS transistor level schema.

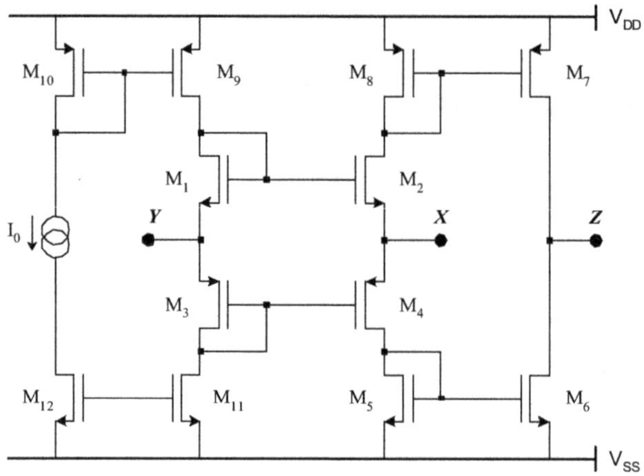

Figure 5. The second generation CMOS current conveyor

Constraints:

- Transistor saturation conditions: all the CCII transistors must operate in the saturation mode. Saturation constraints of each MOSFET were determined. For instance, expression (5) gives constraints on M2 and M8 transistors:

$$\sqrt{\frac{I_0}{K_P W_P / L_P}} \le \frac{V_{DD}}{2} - |V_{TP}| - \sqrt{\frac{I_0}{K_N W_N / L_N}} \tag{5}$$

where I_0 is the bias current, $W_{(N,P)}/L_{(N,P)}$ is the aspect ratio of the corresponding MOS transistor. $K_{(N,P)}$ and V_{TP} are technology parameters. V_{DD} is the DC voltage power supply.

Objective functions:

In order to present simplified expressions of the objective functions, all NMOS transistors were supposed to have the same size. Ditto for the PMOS transistors.

- R_X: the value of the X-port input parasitic resistance has to be minimized,
- f_{chi}: the high current cut-off frequency has to be maximized.

Symbolic expressions of the objective functions are not given due to their large number of terms.
PSO algorithm was programmed using C++ software. Table 1 gives the algorithm parameters.
Fig. 6 shows Pareto fronts (R_X vs. fci) and optimal variables (W_P vs. W_N) corresponding to the bias current I_0=50µA (6.a, 6.b), 100µA (6.c, 6.d), 150µA (6.e, 6.f), 200µA (6.g, 6.h), 250µA (6.i, 6.j) and 300µA (6.k, 6.l). Where values of L_N, L_P, W_N and W_P are given in µm, I_0 is in µA, R_X in ohms and $fc_{i(min,\ Max)}$ in GHz.
In Fig. 6 clearly appears the high interest of the Pareto front. Indeed, amongst the set of the non-dominated solutions, the designer can choose, always with respect to imposed specifications, its best solution since he can add some other criterion choice, such as Y-port and/or Z-port impedance values, high voltage cut-off frequency, etc.

Fig. 7 shows Spice simulation results performed for both points corresponding to the edge of the Pareto front, for I_0=100µA, where R_{Xmin}=493 ohms, R_{XMax}=787 ohms, fc_{imin}=0.165 GHz and fc_{iMax}=1.696 GHz.

Swarm size	Number of iterations	w	c_1	c_2
20	1000	0.4	1	1

Table 1. The PSO algorithm parameters

Technology	CMOS AMS 0.35 µm
Power voltage supply	V_{SS}=-2.5V, V_{DD}=2.5V

Table 2. SPICE simulation conditions

I_0	W_N	L_N	W_P	L_P	W_N	L_N	W_P	L_P
	R_{Xmin}		fc_{imin}		R_{XMax}		fc_{iMax}	
50	17.21	0.90	28.40	0.50	4.74	0.87	8.40	0.53
	714		0.027		1376		0.866	
100	20.07	0.57	30.00	0.35	7.28	0.55	12.60	0.35
	382		0.059		633		1.802	
150	17.65	0.6	28.53	0.35	10.67	0.59	17.77	0.36
	336		0.078		435		1.721	
200	17.51	0.53	29.55	0.35	12.43	0.53	20.32	0.35
	285		0.090		338		2.017	
250	18.60	0.54	30.00	0.35	15.78	0.55	24.92	0.35
	249		0.097		272		1.940	
300	19.17	0.55	29.81	0.35	17.96	0.54	29.16	0.35
	224		0.107		230		2.042	

Table 3. Pareto trade-off surfaces' boundaries corresponding to some selected results

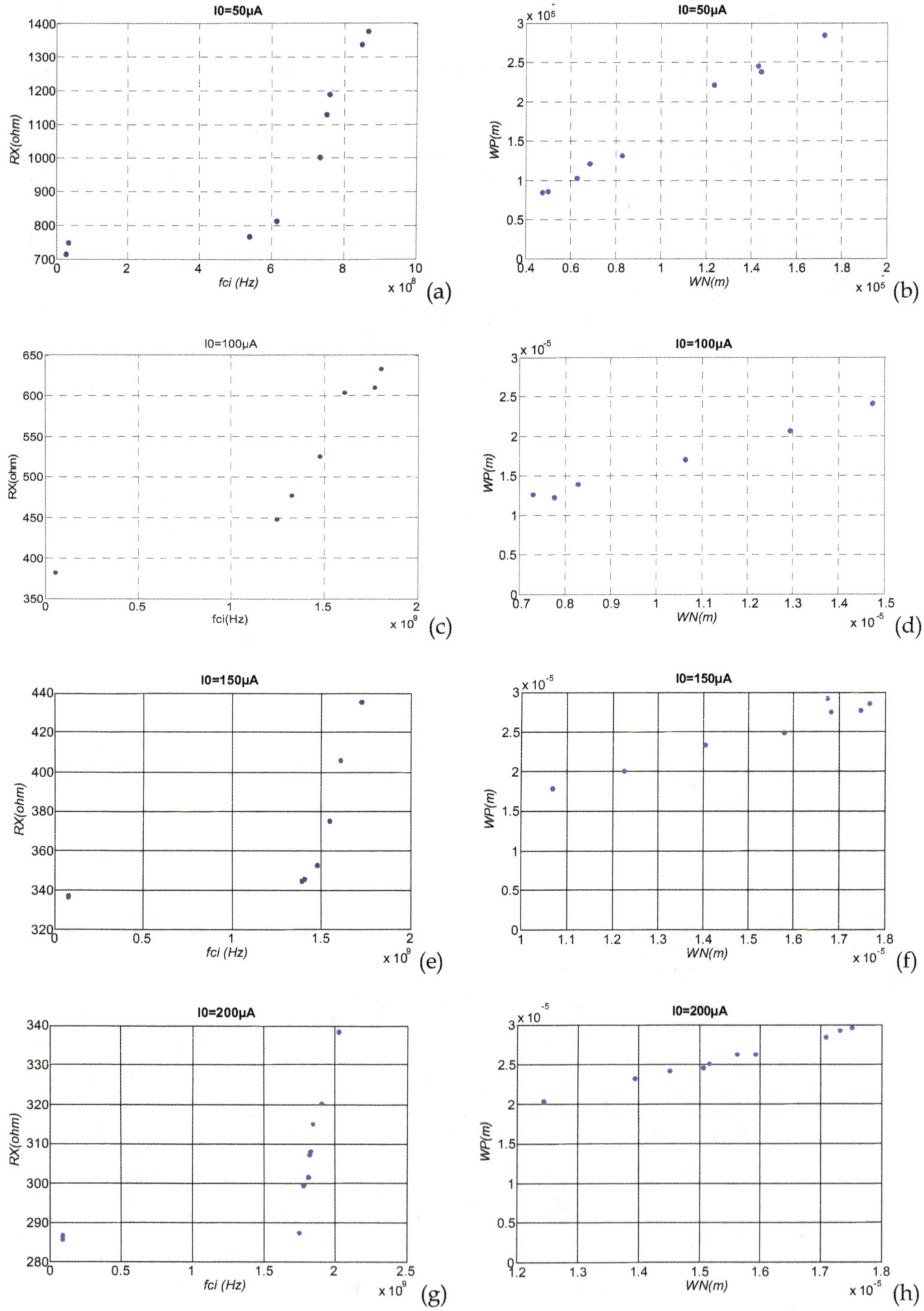
I0=50µA
RX(ohm)
fci (Hz)
x 10^8
(a)
I0=50µA
WP(m)
WN(m)
x 10^5
(b)
I0=100µA
RX(ohm)
fci(Hz)
x 10^9
(c)
I0=100µA
WP(m)
WN(m)
x 10^-5
(d)
I0=150µA
RX(ohm)
fci (Hz)
x 10^9
(e)
I0=150µA
WP(m)
WN(m)
x 10^-5
(f)
I0=200µA
RX(ohm)
fci (Hz)
x 10^9
(g)
I0=200µA
WP(m)
WN(m)
x 10^-5
(h)

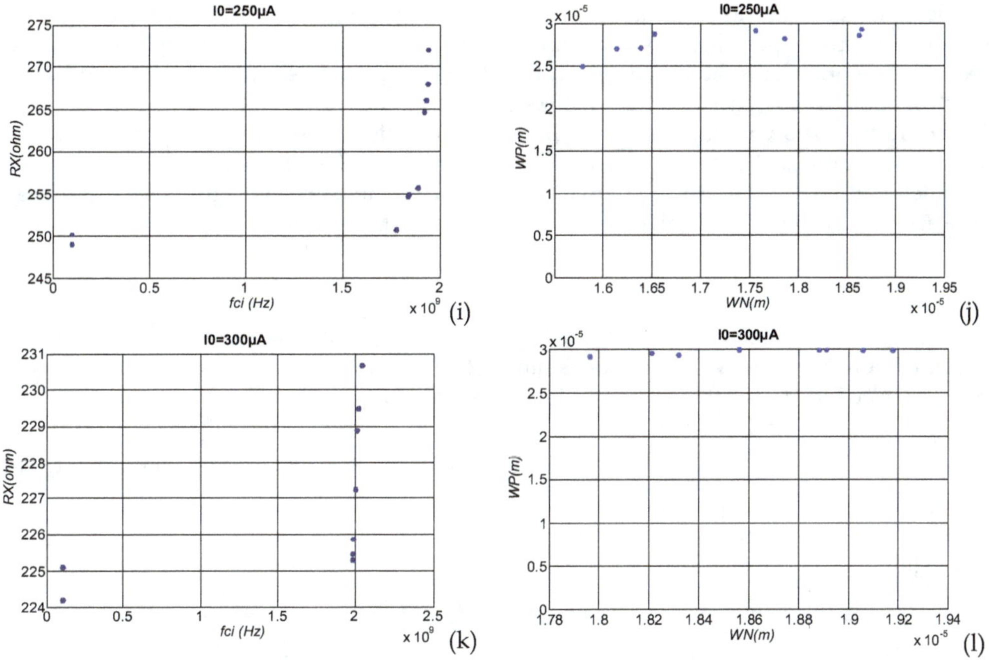

Figure 6. Pareto fronts and the corresponding variables for various bias currents

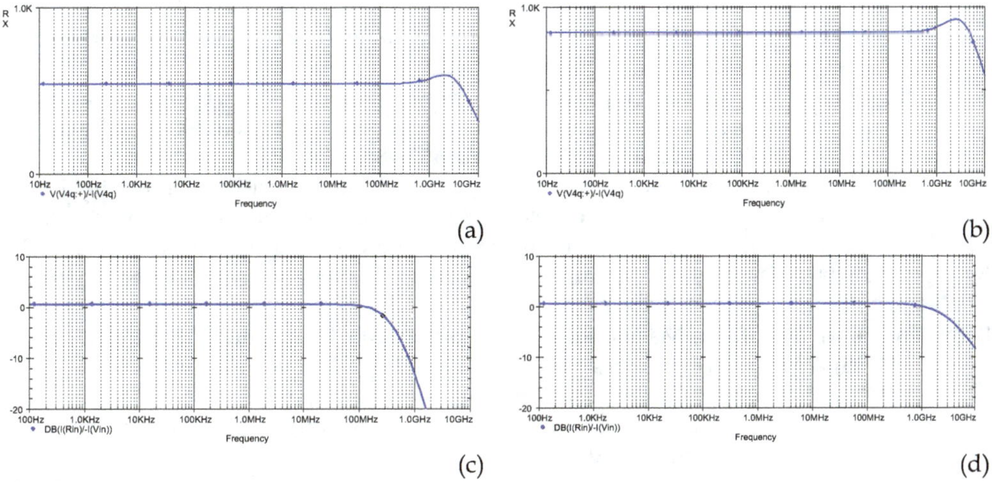

Figure 7. (R_X vs. frequency) Spice simulations

5. Conclusion

The practical applicability and suitability of the particle swarm optimization technique (PSO) to optimize performances of analog circuits were shown in this chapter. An application example was presented. It deals with computing the Pareto trade-off surface in the solution space: parasitic input resistance vs. high current cut-off frequency of a positive second generation current conveyor (CCII+). Optimal parameters (transistors' widths and lengths, and bias current), obtained thanks to the PSO algorithm were used to simulate the CCII+. It was shown that no more than 1000 iterations were necessary for obtaining 'optimal' solutions. Besides, it was also proven that the algorithm doesn't require severe parameter tuning. Some Spice simulations were presented to show the good agreement between the computed (optimized) values and the simulation ones.

6. Appendix

In the analogue sizing process, the optimization problem usually deals with the minimization of several objectives simultaneously. This multi-objective optimization problem leads to trade-off situations where it is only possible to improve one performance at the cost of another. Hence, the resort to the concept of Pareto optimality is necessary.
A vector $\theta = [\theta_1 \cdots \theta_n]^T$ is considered superior to a vector $\psi = [\psi_1 \cdots \psi_n]^T$ if it dominates ψ, i.e., $\theta \prec \psi \Leftrightarrow \underset{i \in \{1,\cdots,n\}}{\forall} (\theta_i \leq \psi_i) \wedge \underset{i \in \{1,\cdots,n\}}{\exists} (\theta_i < \psi_i)$

Accordingly, a performance vector $f^{\bullet}$ is Pareto-optimal if and only if it is non-dominated within the feasible solution space $\Im$, i.e., $\neg \underset{f \in \Im}{\exists} f \prec f^{\bullet}$.

7. References

Aarts, E. & Lenstra, K. (2003) Local search in combinatorial optimization. Princeton University Press.

Banks, A.; Vincent, J. & Anyakoha, C. (2007) A review of particle swarm optimization. Part I: background and development, *Natural Computing Review*, Vol. 6, N. 4 December 2007. DOI 10.1007/s11047-007-9049-5. pp. 467-484.

Basseur, M.; Talbi, E. G.; Nebro, A. & Alba, E. (2006) Metaheuristics for multiobjective combinatorial optimization problems : review and recent issues, report n°5978. National Institute of Research in Informatics and Control (INRIA). September 2006.

Bellman, R. (2003) Dynamic Programming, Princeton University Press, Dover paperback edition.

BenSalem, S; M. Fakhfakh, Masmoudi, D. S., Loulou, M., Loumeau, P. & Masmoudi, N. (2006) A High Performances CMOS CCII and High Frequency Applications, *Journal of Analog Integrated Circuits and Signal Processing*, Springer US, 2006. vol. 49, no. 1.

Bishop, J.M. (1989) Stochastic searching networks, Proceedings of the IEE International Conference on Artificial Neural Networks. pp. 329-331, Oct 1989.

Bonabeau, E.; Dorigo, M. & Theraulaz, G. (1999) Swarm Intelligence: From Natural to Artificial Systems. Oxford University Press.

Chan, F. T. S. & Tiwari, M. K. (2007) Swarm Intelligence: focus on ant and particle swarm optimization, *I-Tech Education and Publishing*. December 2007. ISBN 978-3-902613-09-7.

Clerc, M. Particle swarm optimization (2006), *International Scientific and Technical Encyclopedia*, ISBN-10: 1905209045.

Conn, A. R.; Coulman, P. K.; Haring, R. A.; Morrill, G. L.; Visweswariah, C. (1996), Optimization of custom MOS circuits by transistor sizing, *Proceedings of ICCAD96*. pp. 174-190, November, 1996. San Jose.

Cooren, Y.; Fakhfakh, M. , Loulou, M. & Siarry, P. (2007) Optimizing second generation current conveyors using particle swarm optimization, *Proceedings of the 19th IEEE International Conference on Microelectronics*. December 29-31, 2007. Cairo, Egypt.

Dastidar, T.R.; Chakrabarti, P. P. & Ray, P. (2005). Synthesis System for Analog Circuits Based on Evolutionary Search and Topological Reuse, *IEEE Transactions on Evolutionary Computation*, Vol. 9, No. 2, April 2005. pp. 211-224, ISSN 1089-778X.

Doig, A.H. Land (1960), An automatic method for solving discrete programming problem, *Econometrica*, Vol. 28, pp. 497.

Dorigo, M.; DiCaro G. & Gambardella, L. M. (1999) Ant algorithms for discrete optimization, *Artificial life journal* Vol. 5. pp. 137-172.

Dréo, J.; Petrowski, A.; Siarry, P.; Taillard, E. (2006) *Metaheuristics for Hard Optimization: methods and case studies*, Springer. ISBN: 354023022X.

Fakhfakh M., Loulou M. & Tlelo-Cuautle E. (2007) Synthesis of CCIIs and Design of Simulated CCII based Floating Inductances, *Proceedings of the 14th IEEE International Conference on Electronics, Circuits and Systems*. December 2007. Marrakech, Morocco.

Ferri, G.; Guerrinin, N. & Piccirilli, M. C. (2002) Low voltage current conveyor-based biquad filter, *Proceedings of the IEEE International Symposium on Industrial Electronics*. Vol. 4. pp. 1331-1334.

Glover, F. (1989) Tabu search- part I, *ORSA Journal on Computing*. Vol. 1, N° 3. Summer 1989.

Glover, F. (1990) Tabu search- part II, *ORSA Journal on Computing*. Vol. 2, N° 1. Winter 1990.

Gray, P. R.; Meyer, R. G.(1982) MOS operational amplifier design-a tutorial overview, *IEEE Solid-State Circuits*, Vol. 17, Issue 6, December 1982. pp. 969-982.

Grimbleby, J. B. (2000) Automatic analogue circuit synthesis using genetic algorithms, *Proceedings of the IEE Circuits, Devices and Systems*, December 2000, Vol. 147, Issue: 6. pp: 319-323

Kennedy, J & Eberhart, R. C. (1995) Particle swarm optimization, *Proceedings of the IEEE International Conference On Neural Networks*, pp 1942-1948, WA, Australia.

Kirkpatrick, S.; Gelatt, C. D. & Vecchi, M. P. (1983) Optimization by simulated annealing, *Journal of Science*, 220:671-220-680.

Medeiro, F.; Rodríguez-Macías, R.; Fernández, F.V.; Domínguez-astro, R.; Huertas, J.L.& Rodríguez-Vázquez, A. Global Design of Analog Cells Using Statistical Optimization Techniques, *Analog Integrated Circuits and Signal Processing*, Vol. 6, No. 3, November 1994. pp. 179-195, ISSN 0925-1030.

Nelder, J.A. & Mead, R. (1965) A simplex method for function optimization, *Computer journal*, Vol. 7, pp. 308-313.

Rajput, S. S. & Jamuar, S. S. (2007) Advanced applications of current conveyors: a tutorial, *Journal of active and passive electronic devices*, Vol 2, pp 143-164.

Schmid, H. (2000) Approximating the Universal Active Element, *IEEE Transactions on Circuits and Systems-II: Analog and Digital Signal Processing*, vol. 47, N°. 11, Nov. 2000.

Scniederjans, M.J.(1995) Goal Programming methodology and applications, Kluwer Publishers.

Sedra, A.& Smith, K. C. (1970) A second generation current conveyor and its applications, *IEEE Transactions on Circuit Theory*, pp. 132-134, February 1970.

Siarry(a), P.; Berthiau, G.; Durdin, F. & Haussy, J. (1997) Enhanced simulated annealing for globally minimizing functions of many-continuous Variables, *ACM Transactions on Mathematical Software*, Vol. 23, N°. June 1997. pp 209-228.

Siarry(b), P.; Michalewicz, Z. (2007) Advances in Metaheuristics for Hard Optimization, Springer; December 2007. ISBN 3540729593.

Smith, K. C. & Sedra, A. (1968) The current conveyor-a new circuit building block, *Proceedings of the. IEEE (Letters)*, vol 56, pp. 1368-1369, August 1968.

E. G. Talbi, E. G. (2002) A Taxonomy of Hybrid Metaheuristics, *Journal of Heuristics*, N°8 pp 541-564.

Tlelo-Cuautle, E.; Duarte-Villaseñor, M.A.(2008), Evolutionary electronics: automatic synthesis of analog circuits by GAs, In *Success in Evolutionary Computation, Series: Studies in Computational Intelligence*, Yang, A.; Shan, Y.& Bui, L. T. (Eds.), pp. 165-187, Vol. 92, ISBN: 978-3-540-76285-0, Springer-Verlag, Berlin.

Toumazou, C.& Lidgey, F. J. (1993), In Analog IC Design: The current mode approach, Haigh, D. G. (Ed.), *IEEE circuit and systems series* 2.

Tulunay, G.& Balkir, S. (2004) A Compact Optimization Methodology for Single Ended LNA, *Proceedings of the IEEE International Symposium on Circuits and Systems*, ISCAS 2004.

15

Application of Particle Swarm Optimization Algorithm in Smart Antenna Array Systems

May M.M. Wagih and Hassan M. Elkamchouchi
Alexandria University, Faculty of Engineering
Egypt

1. Introduction

In wireless applications the antenna pattern is shaped so as to cancel interfering signals (placing nulls) and produce or steer a strong beam towards the wanted signal according to signal direction of arrival (DOA). Such antenna system is called smart antenna array.
This chapter presents the efficiency of Particle Swarm Optimization algorithm (PSO) compared to Genetic algorithm (GA) in solving antenna array pattern synthesis problem. Also PSO is applied to determine optimal antenna elements feed that provide null (minimum power) in the directions of the interfering signals while to maximize of radiation in the direction of the useful signal. Application for PSO algorithm in Direct Data Domain Least Squares (D3LS) approach that is used to estimate incoming signal is illustrated.
Due to environment changing the target goal is changing so modification in the algorithm is proposed to provide optimal solution for varying real time target (to track the desired users and reject interference sources). The problem is formulated and solved by means of the proposed algorithm. Examples are simulated to demonstrate the effectiveness and the design flexibility of PSO in the framework of electromagnetic synthesis of linear arrays.

2. Smart Antenna Array System Overview

The ability to communicate with people on the move has evolved remarkably since Marconi first demonstrated radio's ability to provide continuous contact with ships sailing the English Channel in 1897. There onwards, new wireless methods and services have been adopted. Smart antenna system represents one of the valuable parts that support the increasing requirement and needs to higher quality wireless services.
Smart antenna systems processes signals arriving from different directions to detect (estimate) desired signal direction of arrival DOA. Biased on the estimated DOA the beamformer optimize antenna elements weights such that the radiation pattern of the antenna array is adjusted to minimize a certain error function or to maximize a certain reward function derived by the adaptive algorithm. Figure 1. Presents block diagram for Smart antenna system. Smart antenna processing core is represented in three areas the adaptive algorithms the DOA estimation algorithm and the beamformer control.
One of the simplest geometries for an array is a linear array in which the centers of the antenna elements are aligned along a straight line. For simplicity consider the uniformly

spaced linear array of N elements and that there is M signals received. We assume that K samples are observed by the array then output vector $\mathbf{X}(n)$ is

$$\mathbf{X}(n) = \mathbf{A}(\theta)\mathbf{S}(n) + \mathbf{O}(n), \qquad n = 1,2,.... \tag{1}$$

$\mathbf{X}(n)$ is $(N \times K)$ matrix of array output signals at any given instant (sampling time) n, $\mathbf{A}(\theta)$ is $(N \times M)$ steering matrix, $\mathbf{S}(n)$ is $(M \times K)$ signal matrix, $\mathbf{O}(n)$ is noise matrix. The array steering matrix (array manifold) $\mathbf{A}(\theta)$ is

$$\mathbf{A}(\theta) = [\mathbf{a}(\theta_1), \mathbf{a}(\theta_2), \cdots\cdots \mathbf{a}(\theta_{\mathrm{N}})] \tag{2}$$

Where

$$\mathbf{a}(\theta_i) = \left[1, \exp\left(\frac{2\pi d \sin\theta_i}{\lambda}\right), \cdots\cdots \exp\left(\frac{2\pi d(\mathrm{N}-1)\sin\theta_{\mathrm{M}}}{\lambda}\right)\right], \tag{3}$$

$$i = 1,2, \ldots.\mathrm{M}$$

$\mathbf{a}(\theta_i)$ is the response of the linear array to the ith source arriving from direction (θ_i). The array manifold is defined as the one-dimensional manifold composed of all the steering vectors as θ ranges over all possible angles i.e. $\theta \in [0, 2\pi]$.

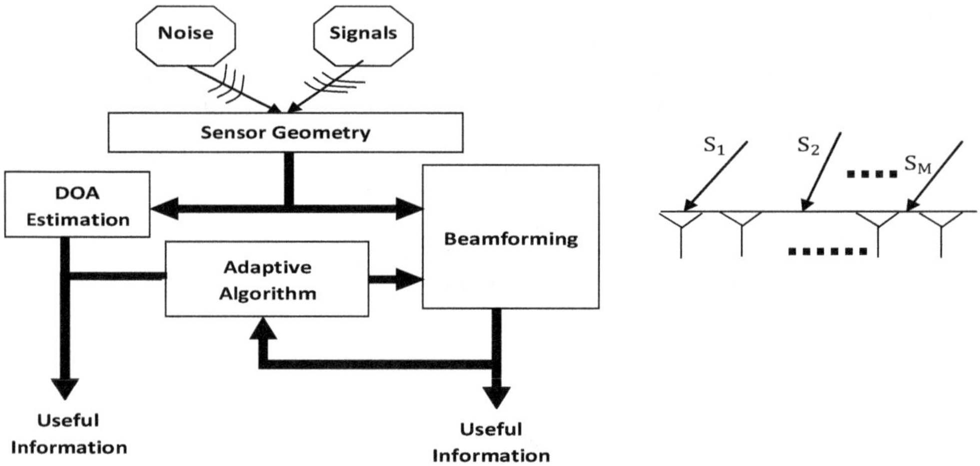

Figure 1. Block diagram of smart antenna array system and linear array signal model

The array manifold used to calibrate the array for direction finding estimation. Each element output is multiplied by a complex weight w_i^*, suggested by the adaptive algorithm then the beamformer update the phase and amplitude relation between the branches, and sum them to give information signal $\mathbf{Y}$(n)

$$\mathbf{Y}(n) = \mathbf{W}(n)^T\mathbf{X}(n) \tag{4}$$

$$\mathbf{W} = [w_1 \quad w_2 \ w_3 \ \ldots w_N]$$

3. PSO for Smart Antenna System

The smart antenna changes their directional pattern with the help of few adjustable parameters in according to the estimation and analysis to received signal, environment and pre-known information to improve the performance and capacity of the system. A promising way for the determination of a suitable parameter configuration for the antenna is the application of heuristic optimization procedures.

Pattern synthesis problem (beamforming) is continuous varying target real time problem that needs fast optimal solution to adjust array pattern and support for the service required. Also the controlling parameters are limited due to practical design and cost aspects. Consequently Enhancement to PSO algorithm is proposed to support for these two major needs.

3.1 PSO and Dynamic Real Environment Optimization

For real time dynamic environment problem the goal value changes, original PSO algorithm has no method for detecting this change and the particles are still influenced by their memories of the original goal position. If the change in the goal is small, this problem is self-correcting. Subsequent fitness evaluations will result in positions closer to the new goal location replacing earlier position **X** vectors, and the swarm should follow, and eventually intersect the moving goal.

However, if the movement of the goal is more pronounced, it moves too far from the swarm for subsequent fitness evaluations to return values better than the current personal best $\mathbf{P}_i^t$ vector, and the particles do not track the moving goal. A proposed attempt to rectify this problem by having the particles periodically replaces their $\mathbf{P}_i^t$ vector with their current $\mathbf{X}_i^t$ vector, thus "forgetting" their experiences to that point. This differs from a restart, in that the particles, in retaining their current location, have retained the profits from their previous experiences, but are forced to redefine their relationship to the goal at that point. Figure 2 present flow chart for the proposed PSO algorithm to support for varying dynamic target optimization problem.

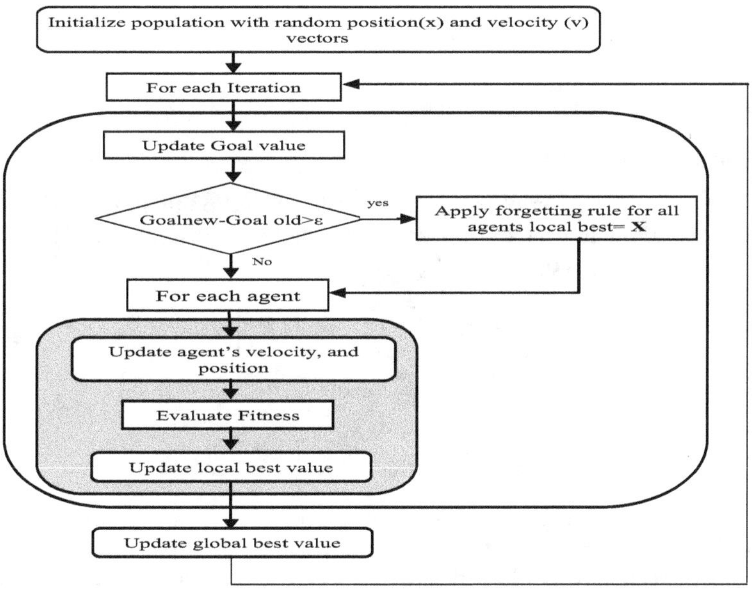

Figure 2. Dynamic Particle Swarm Algorithm

3.2 PSO and bounded search space

Constraint is usually set to the array parameters these constraints may be spatial, for example, that the interelement spacing be greater than a prescribed value or that the element positions be within specified limits. Other type of design constraint is the excitation where it may require that the elements feed is phase only or amplitude and that the current-taper ratio be less than or equal to a prescribed value.

Introducing constraint to the PSO will decrease degree of freedom. Search time will also increase if the concept of accept and after the each particle movement for each iteration according to boundaries. However, if we can convert the problem to an unconstrained one initially through using suitable transformations of the constraint parameter this will eliminate time lost in explore and probability of rejecting the particle movement. Illustration for such solution will be clear in next section while simulation.

4. PSO use for pattern synthesis

This section objects to reformulate and define antenna array adaptive beamforming in term of an optimization problem. Problem Search Space represented by array pattern controlling parameters is identified. Fitness function that measures the deviation of the optimal proposed solution from the target is defined.

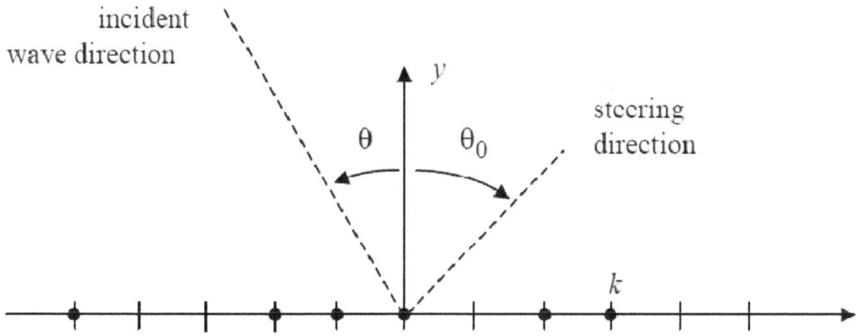

Figure 3.Linear array geometry

Let us consider the linear array of M non-uniformly spaced point source isotropic elements located along a straight line at the positions x_k , where $k = 0, \ldots, M-1$. The beam pattern function $P(u)$ of the array, is defined as follows,

$$P(u) = \left|\sum_{k=0}^{M-1} w_k e^{j\frac{2\pi}{\lambda}x_k u}\right|$$

$$w_k = \propto_k \exp(j\beta_k)$$

$$p(u) = \left|\sum_{k=0}^{M-1} \propto_k e^{j(\frac{2\pi}{\lambda}x_k u+\beta_k)}\right| \tag{5}$$

Where w_k is the weight coefficient of the kth element, λ is the background wavelength, $u = \sin\theta - \sin\theta_o$, being θ and θ_o the incident angle of the impinging plane wave and the steering angle of the array, respectively. In order to generate a beam pattern (BP) that attain specific characteristics e.g., sidelobes level (SLL) lower than a fixed threshold or reproduces a

desired shape $P_{dB}^{ref}(u)$, initially we have to identify the array designing parameters and their boundaries i.e. The particle search space in PSO algorithm.
let vector ζ be defined as follow,

$$\zeta = [M, x_0, \dots x_{M-1}; w_0 \dots . w_{M-1}; D]^T;$$

Where, M is number of array elements, $[x_0, \dots x_{M-1}]$ is array elements spacing vector, $[w_0 \dots . w_{M-1}]$ is array elements feed vector generally represented as $w_k = \alpha_k \, exp\,(j\beta_k)$, finally D is array length. ζ boundary limits has to be taken in account when solving the problem to facilitate practical and cost design needs.
Then a quantized measure for the solution distance from the target required should be defined, this value will be function of the search space parameter vector ζ. Generally for antenna array pattern synthesis most of the well known target consideration is the main beam f_{MB}, total pattern f_{Bp}, sidelobe level f_{SLL}, number, location and width of nulls f_{nul}, number of array elements f_N then we can us define global antenna array fitness function f, as follows:

$$f(\zeta) = \frac{1}{c_1 f_{BP}(\zeta) + c_2 f_{MB}(\zeta) + c_3 f_{SLL}(\zeta) + c_4 f_{nul}(\zeta) + c_5 f_N(\bar{\zeta})} \tag{6}$$

Where

$$f_{BP}(\zeta) = \int_{u \in B} \left(P_{dB}(u)/Q - P_{dB}^{ref}(u) \right) du \tag{7}$$

$$f_{MB}(\zeta) = \sum_{i=1}^{mb} \left(\int_{u \in MB} \left(P_{dB}(u)/Q - P_{dB}^{ref}(u) \right) du \right) \tag{8}$$

$$f_{SLL}(\zeta) = \frac{Q}{max\{P_{dB}(u)\}} \qquad for\ u_{start} \le u \le 1 \tag{9}$$

$$f_{nul}(\zeta) = \sum_{i=1}^{nl} \left(\int_{u \in BN_i} \left(P_{dB}(u)/Q - P_{dB}^{ref}(u) \right) du \right)$$

$$BN_i = u_{nul_i} \pm 0.5 \Delta u_{nul_i} \tag{10}$$

$$f_N(\zeta) = M; \tag{11}$$

Where u_{start} being a value that allows excluding the main lobe from the calculation of the SLL. Moreover, Q is a normalizing constant, B represents visible region; while $P_{dB}^{ref}(u)$ represents the desired BP shape. MB represents the range of values covering the Main beam, mb number of beams in the pattern, BN corresponds to the nulls locations and nl is number of nulls required. Finally, c_i are coefficients that identify each criteria value.
It is often necessary to impose a constraint on the interelement spacing to minimize the mutual coupling effects. For and array with an even number of elements the constraint may be expressed as follow

$$x_1 \ge \frac{d}{2}, \qquad x_i - x_{i-1} \ge d \qquad i = 2,3, \dots \dots M \tag{12}$$

The above constrain can be represented using the following transformation:

$$x_1 = \frac{d}{2} + (\acute{x}_1)^2$$

$$x_2 = \left(\frac{d}{2} + d\right) + (\acute{x}_1)^2 + (\acute{x}_2)^2$$

Generally

$$(x_i) = \left(i - \frac{1}{2}\right)d + \sum_{k=1}^{i} (x'_k)^2 \ , \qquad i = 1,2, \dots . M$$

For odd elements number array

$$(x_i) = (i-1)d + \sum_{k=1}^{i-1} (x'_k)^2 \ , \qquad i = 2, \dots . (M-1)/2 \tag{13}$$

Solving using equation 10 allows minimization to be carried out with the new primed variables, and it is readily seen that the constraints are always satisfied.
Another type of constraint on spacing's usually imposed is the one requiring the elements to lie within a specified range mainly required to avoid unacceptable practical array dimensions. Stated mathematically in the following form:

$$a_i \leq x_i \leq b_i \qquad\qquad i = 1,2, \dots . M \tag{14}$$

the transformation to be used in this case is

$$x_i = a_i + (b_i - a_i)\, sin^2\, x'_i \tag{15}$$

It is sometimes necessary to constrain the current taper to be within specified limits. That is,

$$I_i \leq I \pm C, \qquad\qquad i = 1,2, \dots . \tag{16}$$

It is easily verified that the transformation of the form in equation (15) will transform the constrained space into an unconstrained one

$$I_i = I + C\, sin\, I'_i \tag{17}$$

Next section will investigate the efficiency of the PSO for solving linear array configuration compared to other algorithms.

4.1 PSO and GA for Pattern Synthesis

To validate the PSO approach, initially we apply PSO, to find the optimized element weight to achieve the Chebyshev pattern for 10 equispaced isotropic elements with λ/2 interelement spacing antenna array of minimum SLL of 26dB, and compare its performance to GA, for solving the same problem. The sample points, are chosen 300 equally distributed points over u on a personal computer with a Pentium IV processor running at 1GHz. The target beam will be P_{dB}^{ref}

$$P_{dB}^{ref} = 2.79\, cos\, u + 2.49\, cos\, 3u - 0.97\, cos\, 5u + 1.35\, cos\, 7u + cos\, 9u$$

We consider 10 elements symmetric array with amplitude excitation only i.e. $\beta_i = 0$ then

$$\zeta = \left[w_0 \dots . w_{\left(\frac{M}{2}\right)-1}\right]^T ; \quad M = 10$$

$$w_{k+(M/2)} = w_{(M/2)-(k+1)} \qquad k = 0, \dots M/2$$

$$f_{Bp}(\zeta) = 2 * \int_{u \in B} \left(\frac{P_{dB}(u)}{Q} - P_{dB}^{ref}(u) \right) du\, ; \qquad for\ 0 \leq u \leq 1$$

$$f_{SLL}(\zeta) = \frac{26}{max_{u_{start} \le u \le 1}\{P_{dB}(u)\}} \qquad u_{start} = 0.25$$

$$(\zeta) = \frac{1}{f_{SLL}(\bar{\zeta}) + f_{BP}(\bar{\zeta})}$$

Figure 4 presents the output pattern explored over the optimization process by one particle until it reaches the optimum solution. Corresponding proposed elements weigh for these local minima is as listed in Table 1.

Iteration No	w_0, w_9	w_1, w_8	w_2, w_7	w_3, w_6	w_4, w_5	Max. SLL dB
P1 (5)	0.3292	0.5337	0.7030	0.9883	1	-20
P2 (30)	0.3543	0.3243	0.5679	1	0.7044	-15
P3 (78)	0.3521	0.4688	0.7158	0.8378	1	-12
P4 (122)	0.3574	0.4850	0.7055	0.8921	1	-26

Table 1. Optimum proposed weight corresponding to one particle

Figure 4, 5 shows behavior of the fitness values for solutions explored versus the number of iterations for one particle. Dotted curve represents the gbest fitness value where it intersects with the particles. Note that although the particle has achieved good fitness value in its exploring journey it was not trapped at these local minima at P1, P2, P3, P4.

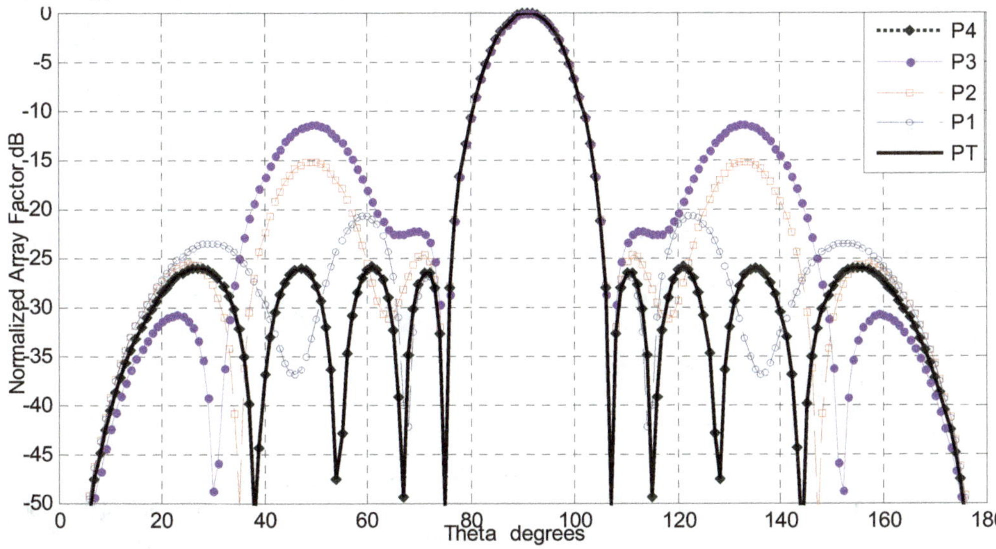

Figure 4. Explored solution for one particle at iteration 5, 30, 80, 120 compared to target pattern

Figure 6 present comparisons between the fitness error per iteration for GA and PSO algorithms solving the above problem with same initial random feed using PSO and Genetic algorithm. It can be noticed the performance difference in reaching optimum solution is not big only difference comes for the time per iteration in each algorithm. According to output in Table 1 that the optimized proposed element feeds is the same for both algorithms.

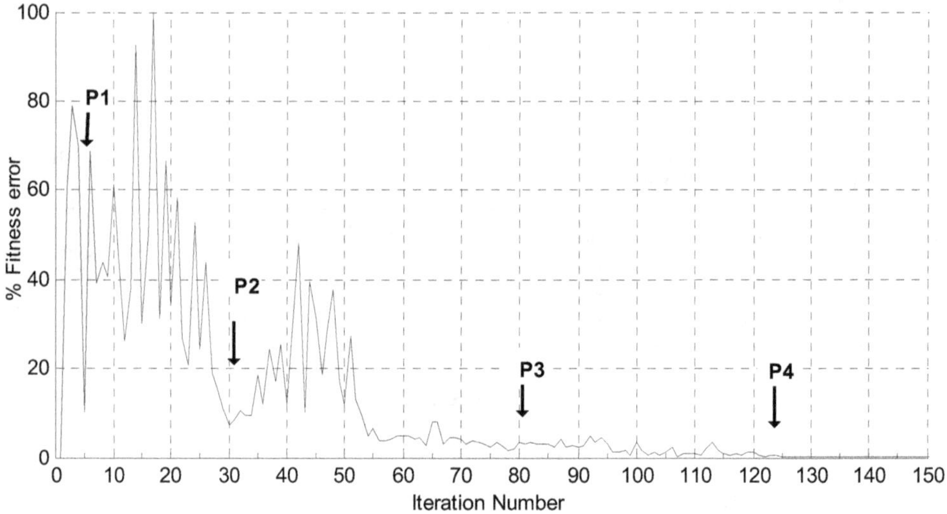

Figure 5. Behavior of the fitness function per iteration for one particle, dotted curve represent behavior of gbest fitness value per iteration

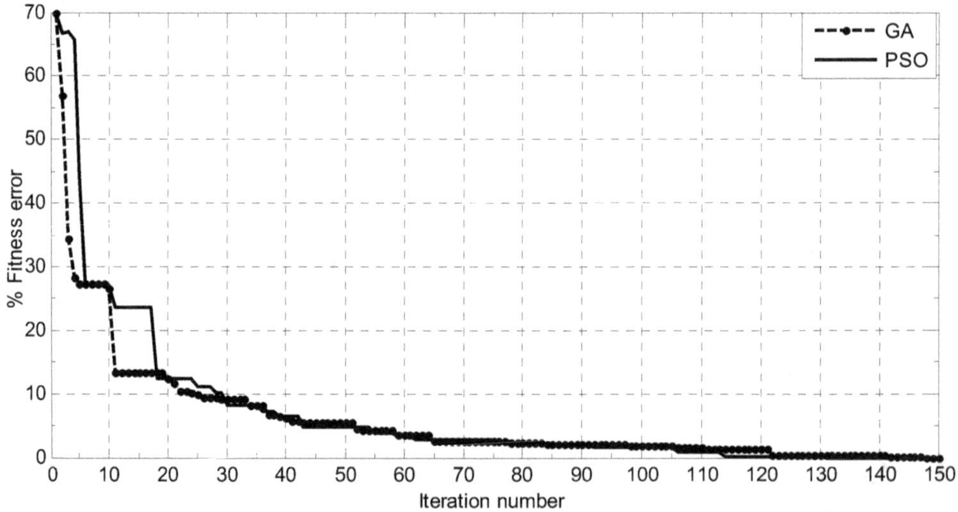

Figure 6. Fitness per iteration behavior for PSO algorithm and GA algorithm

Next section will search the capabilities of the PSO for solving array configuration. A simulation for steering single beam, introducing multiple beams in DOA and introducing nulls in the imposed directions by controlling the excitations of the array elements feed or the elements spacing represented in term of λ. also the adaptive ability of PSO for changing the problem target in runtime is presented such feature is to be useful in digital beamforming.

Algorithm/ Normalized weight	w_0, w_9	w_1, w_8	w_2, w_7	w_3, w_6	w_4, w_5	No. of Iteration	Total time min.
PSO	0.3574	0.4850	0.7055	0.8921	1.000	115	2
GA	0.3563	0.4845	0.7055	0.891	1.000	122	9

Table 2. Optimum proposed weight corresponding to PSO, GA algorithm

4.2 PSO and Pattern Synthesis Phase Control

The phase-only null synthesizing is attractive since in a phased array the required controls are available at no extra cost [Steyskal, H.,1986]. This section will illustrate different scenarios for pattern shaping using PSO to search suitable phase feed to fullfill-required pattern. Initially consider it is required to Introduce single null at direction θ =50° and SLL<30dB with same mainbeam. PSO evaluated element weighting which fulfilled the requirements of the design using fitness function equation 6.
Figure 7, shows the output pattern after 200, iteration notice that the SLL criterion is not achieved.

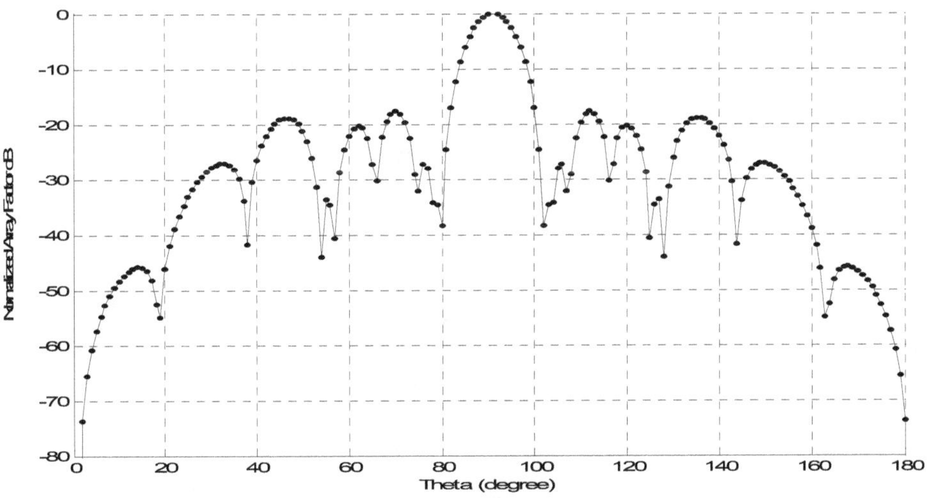

Figure 7. Pattern proposed after 200 iteration for null at 50°

Now let us consider the target is moved. Assume it is required to steer the mainbeam to be at θ =110° and presence of interference at θ =150°. PSO evaluated antenna array elements' phase which fulfill these requirements of the design output proposed pattern as Figure 8a shows the output pattern after 50, iteration as can be notices although that the SLL< 20dB was not achieved as the maximum level is 18dB. Assume surrounding environment is stable so the algorithm is to continue search for better feeding solution Figure 8b shows the proposed pattern corresponding after 500 iteration maximum SLL of -22dB was achieved and also the null width and is increased. Figure 9 shows the total fitness value per iteration curve corresponding to Figure 7 and Figure 8.

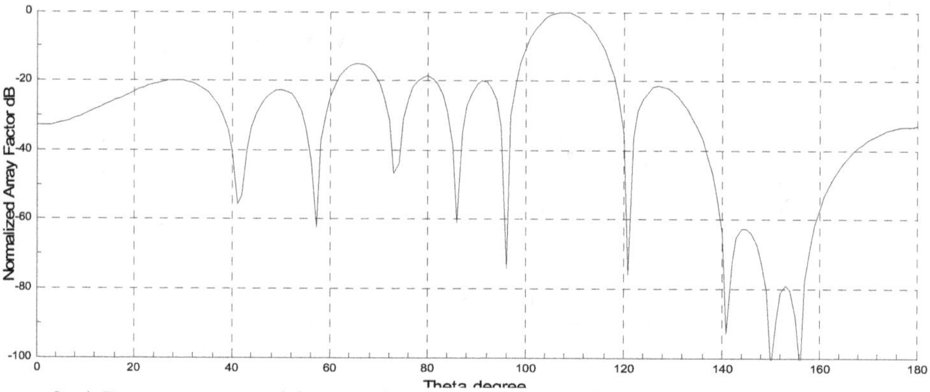

Figure 8. a) Pattern proposed for mainbeam steered to 110° and null at 150° after 50 iterations

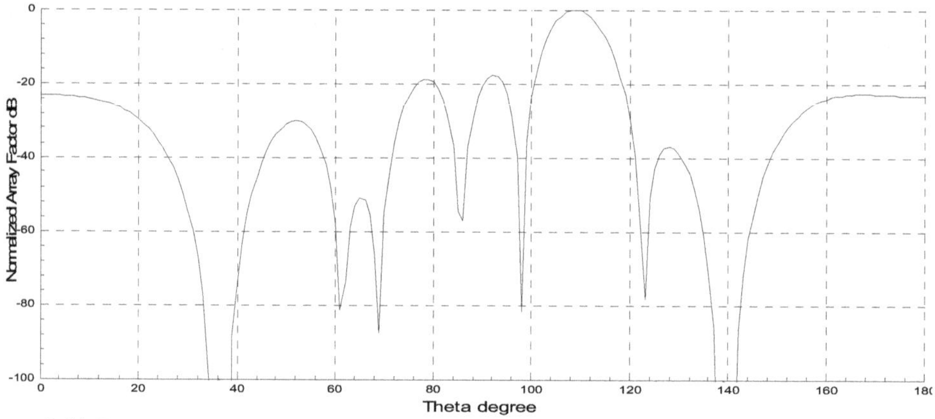

Figure 8. b) Pattern proposed for mainbeam steered to 110° and null at 150° after 500 iteration

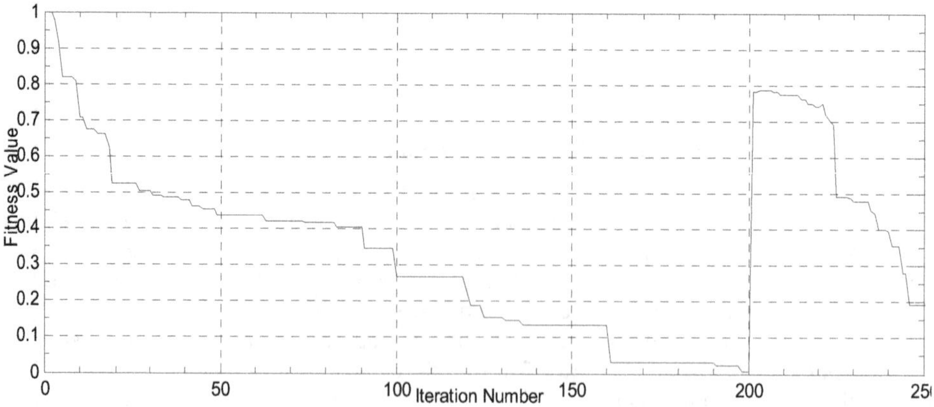

Figure 9. Fitness per iteration curve corresponding to figure 7, 8

4.3 PSO and Pattern Synthesis Phase – Position Control

The phase-only synthesis with equal element spacing requires a large number of elements compared to the amplitude only arrays. Controlling the inter element space and elements phases feed we can have the potential to circumvent this design challenge. Theoretically, the unequal spacing of antenna elements corresponds to nonuniform sampling of signals in the time domain.[H. Unz, 1960] .

The PSO is applied to search for the optimum element phases and positions of the uniform amplitude linear arrays to achieve target pattern and minimum side lobe level .We only consider symmetric arrays for the next results however same can be applied for non symmetric array. Synthesis of an unequally spaced array is carried out separately for the position-only and the position-phase cases for various limits in the distance between the elements. The number of elements considered for the PSO-based synthesis is 32; hence the number of parameters to be optimized is 16 for the position- only synthesis and 32 for the phase-position synthesis.

The PSO synthesis results of positions and phases for the cases when $d_{max} = 0.6\lambda$ and $d_{max} = \lambda$ array patterns are shown in Figure. 10 and 11, respectively. From Figure 10, we can see that the maximum SLL for the position-phase synthesis is lower than that for the position-only synthesis. In Figure 11 When $d_{max} = \lambda$, the maximum SLL of the position-phase synthesis and position-only synthesis is 23.34 and 22.53 dB, respectively

For the case $d_{max} = \lambda$, The time taken to reach -20 dB SLL was about 10 min, and the total time taken for 300 iterations was about 23 min for a swarm of 320 agents. The simulations were carried out on a PC based on an Intel Pentium-IV 3-GHz processor.

We can conclude that for smaller, $d_{max}$the element phases have a larger effect in lowering the SLL of an unequally spaced array with no significant difference in the directivity From Figures 10–11.

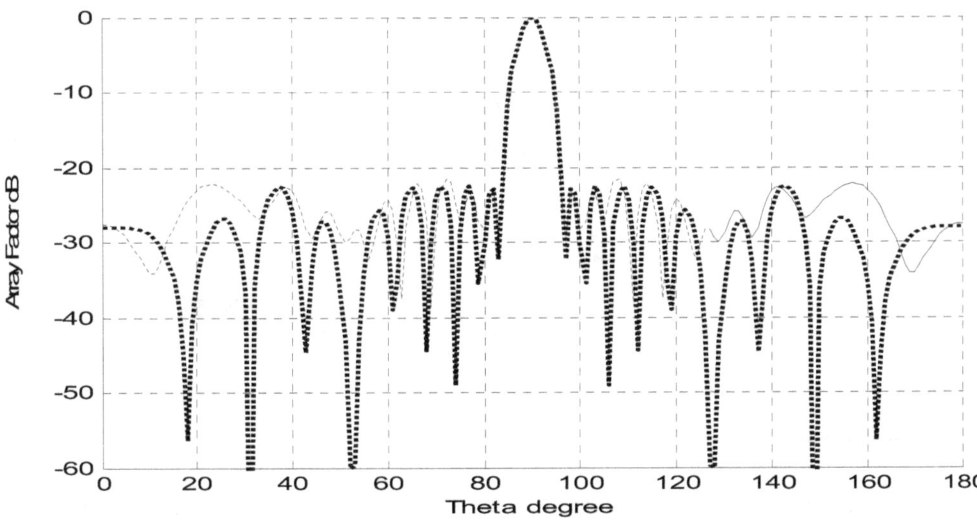

Figure 10. Array patterns for the PSO-based position-only (dashed line) the position-phase (solid line) for $d_{max} = 0.6\lambda$

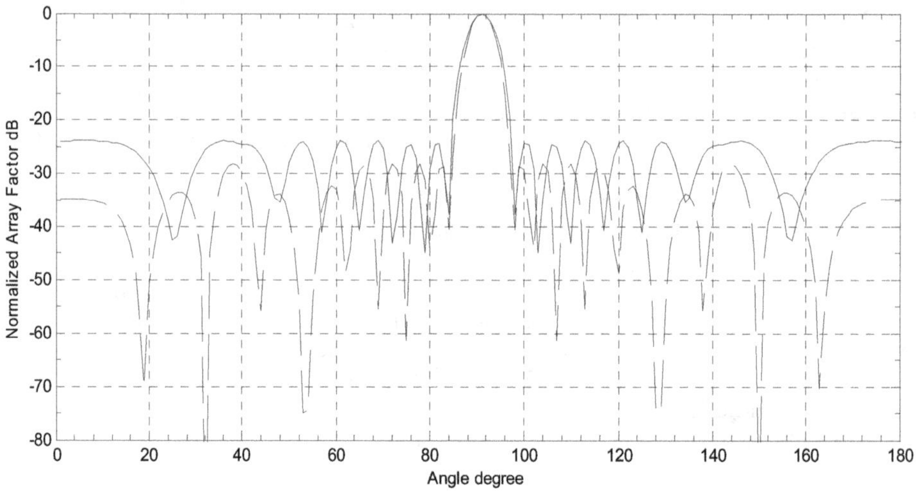

Figure 11. Array patterns for the PSO-based position-only (dashed line) the position-phase (solid line) for $d_{max} = \lambda$

We have seen that the unequally spaced array derived using the position-phase synthesis has lowered SLL compared to that of the unequally spaced arrays derived using the position-only synthesis. Let us consider the PSO-based position-phase synthesis and phase-only synthesis for designing a pencil beam array.

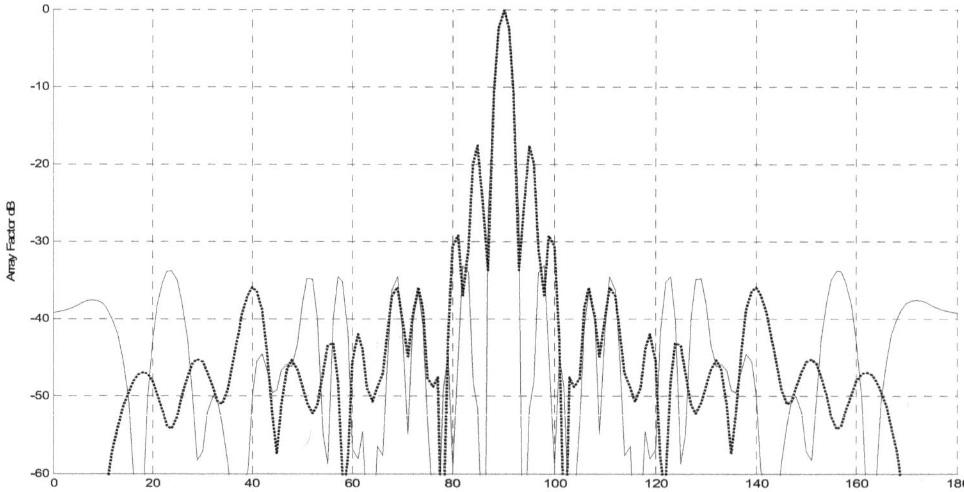

Figure 12. Array patterns for the PSO -based position-phase synthesis (solid line) and the phase-only synthesis (dashed line) of a pencil beam array of 60 elements

The number of elements has to increase to meet beam requirement we consider symmetric array of 60 elements. For the position-phase synthesis, the prior limits assumed in the minimum and maximum distance between the elements are $d_{min} = 0.5\lambda$ and $d_{max} = 0.7\lambda$, respectively. For phase-only synthesis, the uniform distances between the elements are assumed to be 0.5λ. Figure 12 shows the corresponding array patterns shows the phases and positions derived using the PSO-based phase-only synthesis and position-phase synthesis we can see that for the position-phase synthesis, the SLL is lower compared to that of the phase-only synthesis.

5. PSO Application in Smart Antenna Array Signal Estimation

Conventional adaptive beamforming algorithms are based on a stationary environment. Assume that the desired signal and interferers are not correlated. Using statistical theory, one requires several successive snapshots of the data to form a covariance matrix of the interference with independent identically distributed secondary data[B. D. Van, IEEE 1986]. The snapshots accumulation is quite time consuming. Thus when the environment becomes nonstationary, an inaccurate covariance matrix is derived, which results in that the interference cannot be rejected. Therefore the adaptive processing using a single snapshot [Markus E. Ali] is more suitable for a dynamic environment. A direct data domain least squares (D3LS) algorithm [T. K. Sarkar, 2000] has been developed to analyze the received data using a single snapshot.

Although the D3LS algorithm has certain advantages, it has some drawbacks such that the degrees of freedom are limited to nearly half. Furthermore it is shown by simulations that while the jammers can be rejected, the main lobe of the antenna beam pattern is often deviated from the direction of the desired signal and the sidelobe level is relative high.

5.1 Algorithm Formulation

Consider an array composed of N sensors separated by a distance as shown in Figure 1. We assume that narrowband signals consisting of the desired signal plus possibly coherent multipath and jammers with center frequency $f_\circ$ are impinging on the array from various angles, with the constraint. For sake of simplicity, we assume that the incident fields are coplanar and that they are located in the far field of the array.

Each received signal $x_m(k)$ includes additive, zero mean, Gaussian noise. Time is represented by the kth time sample. Thus, for $\mathrm{X(t)} = [x_1(k) \quad x_2(k) \quad x_\mathrm{N}(k)]^\mathrm{T}$

$$x(k) = [\bar{a}(\theta_1)\ \bar{a}(\theta_2) \ldots.\ \bar{a}(\theta_M)]. \begin{bmatrix} s_1(k) \\ s_2(k) \\ \vdots \\ \vdots \\ s_M(k) \end{bmatrix} + \bar{n}(k) \tag{19}$$

$\bar{a}(\theta_i)$ is M-elements array steering vector for the θ_i direction of arrival, λ wavelength and d is the elements interspacing distance. $\bar{s}(k)$ is the vector of incident signals at time k and $\bar{n}(k)$ is noise vector at each array element m, zero mean, variance. Then for

$$\bar{A} = [\bar{a}(\theta_1)\ \bar{a}(\theta_2) \ldots.\ \bar{a}(\theta_D)]_{M\times D}$$

matrix of steering vectors $\bar{a}(\theta_i)$

$$\overline{\mathrm{X}} = \bar{A}.\bar{s}(k) + \bar{n}(k) \tag{20}$$

Thus, each of the D-complex signals arrives at angles θ_i and is intercepted by the M antenna elements. It is assumed the number of arriving signals D < M. It is understood that the arriving signals are time varying and thus our calculations are based upon time snapshots of

the incoming signal. Obviously if the transmitters are moving, the matrix of steering vectors is changing with time and the corresponding arrival angles are changing.
Let S_n be the complex voltage induced in the nth array element at a particular instance of time due to a signal of unity amplitude coming from a direction θ_s,

$$S_n = exp\left[j2\pi\left\{\frac{(n-1)d}{\lambda}sin(\theta_s)\right\}\right] \tag{21}$$

Let x_n be the complex voltages that are measured at the nth element due to the actual signal, jammers and thermal noise

$$x_n = \alpha_s S_n + \text{Interference} + \text{Noise} \tag{22}$$

$$x_n = \alpha_s S_n + \sum_{p=1}^{D-1}\left(A_p exp\left(j\frac{2\pi(n-1)d}{\lambda}sin(\theta_p)\right)\right) + n_n \tag{23}$$

Where α_s denotes the complex amplitude of the desired SOI, A_p and θ_p are the amplitude and direction of arrival of the p jammer signal, n_n is the thermal noise at the nth element. There are D jammers and $D < M - 1/2$. With S_n and X_n (n=0,...,M) the known received signal data, one can construct the matrix **X** and **S** such that

$$\boldsymbol{X} = \begin{bmatrix} x_0 & x_1 & \cdots & x_L \\ x_1 & x_2 & \cdots & x_{L+1} \\ \vdots & \vdots & \cdots & \vdots \\ x_L & x_{L+1} & \cdots & x_M \end{bmatrix}_{(L+1)\times(L+1)} \qquad \boldsymbol{S} = \begin{bmatrix} s_0 & s_1 & \cdots & s_L \\ s_1 & s_2 & \cdots & s_{L+1} \\ \vdots & \vdots & \cdots & \vdots \\ s_L & s_{L+1} & \cdots & s_M \end{bmatrix}_{(L+1)\times(L+1)} \tag{24}$$

From equations (21) and (23) the matrix $\mathrm{U} = \mathrm{X} - \alpha_s \boldsymbol{S}$, represents the contribution due to signal multipaths, interferes, clutter and thermal noise (i.e., all the undesired components of the signals except SOI). In an adaptive beamforming, the adaptive weight vector w is chosen in such a way that the contribution from the jammers and thermal noise are minimized to enhance the output signal to interference plus noise ratio (SINR). Hence, the following generalized eigenvalue problem is obtained.

$$\mathrm{UW} = (\mathrm{X} - \alpha_s S)W = 0 \tag{25}$$

$$\overline{w} = [w_1 \; w_2 \; .... \; w_N]^T$$

Note that U(1,1) and U(1,2) elements of the interference plus noise matrix, are given by

$$U(1,1) = X_1 - \alpha S_{d1} \tag{26}$$

$$U(1,2) = X_2 - \alpha S_{d2} \tag{27}$$

Where X_1 and X_2 are the voltages received at antenna elements 1 and 2 due to the signal, jammer, clutter and noise where as S_{d1} and S_{d2} are the values of the SOI only at those elements due to a signal of unit strength, let us define Z as follow

$$z = exp\left[j2\pi\left\{\frac{d}{\lambda}sin(\theta_s)\right]\right. \tag{28}$$

Then $U(1,1) - z^{-1}U(1,2)$ contains no component of the desired signal. In general, the same is true for $U(i,j) - z^{-1}U(i,j+1), (i = 1, ..., L+1, j = 1, ..., L)$. Therefore one can form a square matrix F of dimension L+1, generated from U. Therefore, in such way, one can form a reduced rank matrix combined with a constraint that the gain of the subarray is C in the directionθ_s, then one can obtain equation given as follow

$$\begin{bmatrix} Z^0 & Z^1 & \dots & Z^L \\ X_0 - Z^{-1}X_1 & \dots & \dots & X_L - Z^{-1}X_{L+1} \\ \vdots & \vdots & \vdots & \vdots \\ f_3 - f_4 Z^{-1} & \dots & \dots & X_{M-1} - Z^{-1}X_M \end{bmatrix} \begin{bmatrix} W_0 \\ W_1 \\ \vdots \\ W_L \end{bmatrix} = \begin{bmatrix} C \\ 0 \\ 0 \\ 0 \end{bmatrix} \quad (29)$$

To obtain the desired signal component, equation (5.14) is represented as

$$[\mathrm{F}][\mathrm{W}] = [\mathrm{Y}] \quad (30)$$

Using any optimization algorithm to solve equation (30) for, optimum weight vector [W] that provide maximum signal gain through minimizing objective function represented as equation (31)

$$\zeta(\mathrm{W_i}) = \frac{\|[\mathrm{F}][\mathrm{W_i}] - \mathrm{Y}\|}{\|\mathrm{Y}\|} \leq 10^{-6} \quad (31)$$

Consequently SOI the signal component α may be estimated from

$$\alpha = \frac{1}{\mathrm{C}} \sum_{i=0}^{i=L} [\mathrm{W_i X_i}] \quad (32)$$

The algorithm above is referred to as a "forward method" in the literature [8]. [6],[11]. note we can reformulate the problem using the same data to obtain independent estimate for the solution. This can be achieved by two methods:

a. By reversing the data sequence and then complex conjugating each term of that sequence (Backward method)
b. By combining the (forward-backwards method) to double the given data and thereby increase the number of weights (degrees of freedom) significantly over that of either the forward or backward method alone. The number of degrees of freedom can reach to $1 + (N - 1)/1.5$.

to investigate the method let us we consider recovering signal using the previous presented algorithm let us consider a single tone signal with specs as table (3) received by liner array of 10 elements linear array.

	Magnitude in V	Phase	DOA in degree
Signal	1	0	45°
Jammer #1	1.25	0	75°
Jammer # 2	2	0	60°
Jammer #3	0.5	0	0°

Table 3. Incident signal characteristics

The sampling frequency is 10 f; Using PSO algorithm as an optimization tools to solve the optimum $\mathrm{W_i}$ for the objective equation (31) value for each iteration we get

$\mathrm{W_1}$= (1.2996248637+j*0.0724160744), $\mathrm{W_2}$=(0.9415241429+j*-0,3236468668)

$\mathrm{W_3}$=(-0.9898155714+j*-0.1071454180), $\mathrm{W_4}$ = (- 1.2513334352+j*0.3583762104)

Using these weights in equation 32 to get the value of SOI amplitude
The first ten samplings of the signal and the system output are compared as follow

Initial transmitted signal	Estimated signal
1	0.9999-j0.000003154
0.809+j0.5877	0.809+j0.5877
0.309+j0.951	0.309+j0.951
-0.309+j0.951	-0.309+j0951
-0.809+j0.5877	-0.809+j0.5877
-1	-0.90.999+j0000003154
-0.809-j0.587	-0.809+j0.5877
-0.309-j0.951	-0.309-j0.951
0.3090-j0.951	0.309-j0.951
0.809-j0.5877	0.80.90-j0.5877

Table 4. Output estimated signal using D3LS and PSO algorithm as an optimization method

The total CPU time taken for the above results is 1.19 sec. PSO is less computational operations compared to conjugate gradient method.

6. Conclusion

PSO application for solving different numerical problems in smart antenna is illustrated. Improvement is proposed to the algorithm to support the continuous real time varying target problem. Also a solution is proposed to overcome the case of bounded search space through introducing of transformation function. Simulation for different scenarios is solved with the aid of PSO. Synthesis of an adaptive Beamforming using the phase only control where target is dynamic over time has been presented. PSO was introduced to solve position-only and position-phase synthesis, which is a bounded search space problem. Finally an investigation for using PSO to estimate signal amplitude though D3LS approach is presented.

7. References

H. Unz, Linear arrays with arbitrarily distributed elements, *IEEE Trans. Antennas Propagat.*, vol. AP-8, pp. 222–223, Mar. 1960

B. D. Van Veen and K. M. Buckley, Beamforming: a versatile approach to spatial filtering, *IEEE SSP Mag., Apr.1988, pp. 4-24*

Markus E. Ali and Franz Schreib, Adaptive Single Snapshot Beamforming: A New Concept for the Rejection of Nonstationary and Coherent Interferers, *IEEE Trans. On Signal Processing*, Vol. 40, No. 12, Dec.1992, pp. 3055-3058.

Steyskal, H., R. A. Shore, and R. L. Haupt, Methods for null control and their effects on radiation pattern, *IEEE Trans. On Antenna and Propagation*, Vol. 34, 404–409, 1986.

2. Steyskal, H., Simple method for pattern nulling by phaseperturbation, *IEEE Trans. on Antenna and Propagation*, Vol. 31,163–166, 1983.

T. K. Sarkar and J. Koh, A Pragmatic Approach to Adaptive Antennas, *IEEE Antennas and Propagation Magazine*, Vol. 42, No. 2, Apr. 2000, pp.39-53

Permissions

All chapters in this book were first published in PSO, by InTech Open; hereby published with permission under the Creative Commons Attribution License or equivalent. Every chapter published in this book has been scrutinized by our experts. Their significance has been extensively debated. The topics covered herein carry significant findings which will fuel the growth of the discipline. They may even be implemented as practical applications or may be referred to as a beginning point for another development.

The contributors of this book come from diverse backgrounds, making this book a truly international effort. This book will bring forth new frontiers with its revolutionizing research information and detailed analysis of the nascent developments around the world.

We would like to thank all the contributing authors for lending their expertise to make the book truly unique. They have played a crucial role in the development of this book. Without their invaluable contributions this book wouldn't have been possible. They have made vital efforts to compile up to date information on the varied aspects of this subject to make this book a valuable addition to the collection of many professionals and students.

This book was conceptualized with the vision of imparting up-to-date information and advanced data in this field. To ensure the same, a matchless editorial board was set up. Every individual on the board went through rigorous rounds of assessment to prove their worth. After which they invested a large part of their time researching and compiling the most relevant data for our readers.

The editorial board has been involved in producing this book since its inception. They have spent rigorous hours researching and exploring the diverse topics which have resulted in the successful publishing of this book. They have passed on their knowledge of decades through this book. To expedite this challenging task, the publisher supported the team at every step. A small team of assistant editors was also appointed to further simplify the editing procedure and attain best results for the readers.

Apart from the editorial board, the designing team has also invested a significant amount of their time in understanding the subject and creating the most relevant covers. They scrutinized every image to scout for the most suitable representation of the subject and create an appropriate cover for the book.

The publishing team has been an ardent support to the editorial, designing and production team. Their endless efforts to recruit the best for this project, has resulted in the accomplishment of this book. They are a veteran in the field of academics and their pool of knowledge is as vast as their experience in printing. Their expertise and guidance has proved useful at every step. Their uncompromising quality standards have made this book an exceptional effort. Their encouragement from time to time has been an inspiration for everyone.

The publisher and the editorial board hope that this book will prove to be a valuable piece of knowledge for researchers, students, practitioners and scholars across the globe.

List of Contributors

Djerou L., Khelil N. and Zerarka A.
Labo de Mathématiques Appliquées, Université Med Khider, Algeria

Batouche M.
Computer Science Department, CCIS-King Saud University, Saudi Arabia

Takeshi Matsui, Masatoshi Sakawa, Kosuke Kato and Koichi Tamada
Hiroshima University, Japan

Tzung-Pei Hong
Department of Computer Science and Information Engineering, National University of Kaohsiung
Department of Computer Science and Engineering, National Sun Yat-sen University

Guo-Neng Shiu
Department of Electrical Engineering, National University of Kaohsiung

Yeong-Chyi Lee
Department of Information Management, Cheng-Shiu University, Taiwan

Andrei Băutu
"Mircea cel Batrân" Naval Academy

Elena Băutu
"Ovidius" University, Romania

Han Huang
School of Software Engineering, South China University of Technology

Zhifeng Hao
College of Mathematical Science, South China University of Technology, P. R. China

M. Fikret Ercan
Singapore Polytechnic School of Electrical and Electronic Engineering, Singapore

Mahamed G.H. Omran
Gulf University for Science and Technology, Kuwait

Chang-Huang Chen
Tungnan University

Jonq-Chin Hwang and Sheng-Nian Yeh
National Taiwan University of Science and Technology Taipei, Taiwan, ROC

Po-Hung Chen
Department of Electrical Engineering, St. John's University, Taiwan

Patrick Marques Ciarelli, Renato A. Krohling and Elias Oliveira
Universidade Federal do Espírito Santo, Brazil

Men-Shen Tsai and Wu-Chang Wu
National Taipei University of Technology Taipei, Taiwan

Omar Ilaya and Cees Bil
RMIT University, Australia

S.G. Ponnambalam
Monash University, Malaysia

N. Jawahar
Thiagarajar College of Engineering, India

S. Chandrasekaran
S R M V Polytechnic College, India

Mourad Fakhfakh and Mourad Loulou
University of Sfax, Tunisia

Yann Cooren and Patrick Siarry
University of Paris 12, France

May M.M. Wagih and Hassan M. Elkamchouchi
Alexandria University, Faculty of Engineering, Egypt

Index

www.ingramcontent.com/pod-product-compliance
Lightning Source LLC
LaVergne TN
LVHW051046220926
848485LV00001BA/25
* 9 7 8 1 6 3 9 8 9 0 2 4 8 *